URBAN STRUCTURE OF HILL TOWNS

Case of Himachal Pradesh, India

Edited By

Dr. Puneet Sharma

© 2021, Dr. Puneet Sharma

ISBN 978-1-6671-2295-3
Lulu.com

This work is licensed under a Standard Copyright License. All rights reserved.

DEDICATION

Dedicated to my mother....

Table of Contents

ABSTRACT I

List of illustrations II

Acknowledgement V

SUSTAINABLE DEVELOPMENT IN HILL TOWNS 1

Introduction 2

    1. Urban form in hills 4
    2. Intrinsic development aspects in hills 7
    3. Transportation in hill towns 8
    4. Townscape of hill settlements 10

URBAN HILL SETTLEMENTS OF HIMACHAL PRADESH 11

    Introduction 12

OVERVIEW OF SHIMLA TOWN 14

    Introduction 15

    1. Climatic conditions 16
    2. Topography 18
    3. Tourist inflow 20

    Morphology and urban form 21

    1. Land use and Green cover 21
    2. Demographic character 24
    3. Grain and texture 26
    4. Transportation and movement 27

    Cityscape 30

    1. Landmark and activity generators 30
    2. Vernacular architectural vocabulary 31
    3. Contemporary architectural expression 32

Conclusion 33

OVERVIEW OF MANALI TOWN 35

Introduction 36

    1. Climatic conditions 37
    2. Topography 39
    3. Tourist inflow analysis 41

Morphology and urban form 42

    1. Land use and Green cover 42
    2. Demographic character 44
    3. Grain and texture 45
    4. Transportation and movement 46

Cityscape 47

    1. Landmark and activity generators 47
    2. Vernacular architectural vocabulary 48
    3. Contemporary architectural expression 50

Conclusion 51

OVERVIEW OF DHARMASHALA TOWN 53

Introduction 54

    1. Climatic conditions 55
    2. Topography 57
    3. Tourist inflow 58

Morphology and urban form 59

    1. Land use and Green cover 59
    2. Demographic character 60
    3. Grain and texture 61
    4. Transportation and movement 62

- Cityscape 64
  1. Landmark and activity generators 64
  2. Vernacular architectural vocabulary 65
  3. Contemporary architectural expression 66
- Conclusion 67
- OVERVIEW OF Sujanpur TOWN 68
  - Introduction 69
    1. Climatic conditions 70
    2. Topography 71
    3. Tourist inflow 72
  - Morphology and urban form 73
    1. Land use and Green cover 73
    2. Demographic character 74
    3. Grain and texture 75
    4. Transportation and movement 75
  - Cityscape 76
    1. Landmark and activity generators 76
    2. Vernacular architectural vocabulary 78
    3. Contemporary architectural expression 79
  - Conclusion 80
  - Synthesis 81
  - Bibliography 82
  - ABOUT THE AUTHOR 87

ABSTRACT

Urban design is the art of creating spaces which are public in nature and intended to be used by all. Derived from the idea of civic design at city level presently urban design is considered as most recognized way to deal with urban complexities. This book contains study of few case examples of urban settlements in hill towns and opens a new perspective for urban designers working in these regions. Urban design exploration in such area starts with the basic understanding of topographical, geographical, climatic, cultural and economic conditions.

Urban hill settlements are recognized by their picturesque beauty, vegetation, clean climate and serene environment. The inherent livable qualities have contributed in making such urban settings as preferred destination for peaceful living and leisure tourism. Urban design interventions in Indian hill towns are very limited. This is due lack of awareness and availability of experts in the field. This book is an attempt to set basic methodology for study and analysis so as to derive context based urban design agendas. Study and analysis of existing city structures and urban form in the Hill towns to achieve sustainable development needs an urban design perspective. Such explorations will help in developing urban design guidelines for future development.

LIST OF ILLUSTRATIONS

Figure 1	Map of Himachal Pradesh	14
Figure 2	Location of Shimla, Himachal Pradesh	17
Figure 3	Average temperature in Shimla	20
Figure 4	Average rainfall in Shimla	20
Figure 5	Contour map of Shimla city	22
Figure 6	Tourist inflow per day	23
Figure 7	Land use map of Shimla city	25
Figure 8	Demographic character of Shimla	28
Figure 9	Figure ground map of Shimla	29
Figure 10	Main transit nodes in city	31
Figure 11	Major activity zones in Shimla	33
Figure 12	Town hall Shimla	34
Figure 13	Building with old residential character	34
Figure 14	Sloping roof form in new construction	35
Figure 15	Framed construction for multistory	35
Figure 16	Location map of Kullu, Himachal Pradesh	39
Figure 17	Average rainfall in Manali	41
Figure 18	Average temperature in Manali	41
Figure 19	Contour map of Manali, Himachal Pradesh	43
Figure 20	Tourist inflow per day	44
Figure 21	Land use map of Manali town	46
Figure 22	Demographic character of Manali town	47

Figure 23	Figure ground map of Manali town	48
Figure 24	Main transit routes in city	49
Figure 25	Major activity zones in Manali	50
Figure 26	Blend of traditional and modern	51
Figure 27	Building with old residential character	52
Figure 28	Framed construction for residences	53
Figure 29	Location map of Kangra, Himachal Pradesh	57
Figure 30	Yearly Average temperature range	59
Figure 31	Yearly Average rainfall	59
Figure 32	Contour map of Dharamshala	60
Figure 33	Tourist inflow per month	61
Figure 34	Land use map of Dharamshala town	62
Figure 35	Demographic character of Dharamshala	63
Figure 36	Figure ground map of Dharamshala	64
Figure 37	Main transit routes in Dharamshala	65
Figure 38	Major activity zones in Dharamshala	67
Figure 39	Typical vernacular house near main town	68
Figure 40	Modern RCC construction with sloping roof	69
Figure 41	Location map of Hamirpur HP	72
Figure 42	Average temperature in Sujanpur	73
Figure 43	Average rainfall in Sujanpur	73
Figure 44	Contour map of Sujanpur	74

Figure 45	Tourist inflow per month	75
Figure 46	Land use map of Sujanpur town	76
Figure 47	Demographic character of Sujanpur	77
Figure 48	Figure ground map of Sujanpur	78
Figure 49	Main transit routes in Sujanpur	78
Figure 50	Major activity zones in Sujanpur	79
Figure 51	View of Chowki, Sujanpur Fort	80
Figure 52	Temple architecture of Sujanpur	80
Figure 53	Typical vernacular house near main town	81
Figure 54	Village setting in hill town	81
Figure 55	Modern street with RCC construction	82

ACKNOWLEDGEMENT

First of all, I would like to express my sincere gratitude to Dr. Inderpal Singh, Head, Department of Architecture, NIT Hamirpur for his time-to-time suggestions, continuous support, patient guidance, and enthusiastic encouragement.

I would like to thank my friends and colleagues for continuous contribution during various data collection stages. I would also like to thank students of NIT Hamirpur for providing useful information and helpful suggestions in making the visual survey. This research was not possible without all residents of various cities who gave their precious time during survey.

And most of all, I owe my deepest gratitude to my wife Soniyanka and my daughter Nishita and son Ritvik for their everlasting love and best wishes that remain indelible in my life. Last but not least, I bow my head to my father Dr. Naresh Kumar and "The Supreme Power-God" equally for bestowing upon me the choicest of their blessings by virtue of which I could reach up to this stage of my academic career. Everyone must not have been mentioned, but none is forgotten.

Puneet Sharma

SUSTAINABLE DEVELOPMENT IN HILL TOWNS

INTRODUCTION

Generic nature of hill towns depends upon their unique land form, urban settings, and diverse landscape. Hills contribute to large ecosystem and have crucial role in protecting rich flora and fauna. Some areas have low load bearing capacity prone to landslides and some are ecologically more sensitive, while others are vantage locations with aesthetic qualities/ scenic viewpoints. In hill towns fast rate of urbanization has transformed the city character. Hill towns have transformed into concrete jungles characterized with depleting forest, unchecked construction, narrow and accident-prone roads, encroachments all over in the recent years. The continuous increasing demand for economic development and tourism demands sustainable development approach in hill areas. In India planning and design of hill settlements has faced numerous failures due to lack of contextual approaches. The development approach in hill towns is case specific. We need to look at hill cities differently due to their geographical constraints like steep slopes, landslides, climate, land use pattern and scarcity of buildable land. Land use and functions should cater to both resident population and floating population. Development should be addressed through urban design frame work while integrating issues of hill stability, ecology and energy with aesthetic qualities of nature.

For sustainable development of hill towns environmental, economic, infrastructural and social factors need to be integrated. Basic issues of sustainable development like urban development and improvement in the quality of life in hill towns are related to their environmental context. Providing a safe, clean and healthy future to coming generations are the main challenges that our world is facing today. To fulfill this aim it is very important that our settlements should be sustainable. Hill regions have very sensitive as well as vital ecosystem which should be maintained by adopting sustainable development techniques. In the hill cities sustainable strategy ranges from infrastructure, socio-cultural, contextual to environmental aspects. These strategies for hill towns guide the aspects for growth and urbanization in a decisive manner. Attributes related to hill regions need due consideration in development strategies so as to preserve its vulnerable ecosystem. Their understanding is a prerequisite for formulation of sustainable strategies. Development approaches must have sustainable policy, transport management and proper awareness among citizens and tourists.

Safety against earthquake is a serious concern for sustainable development in hill regions. Vernacular construction practices like dhajji wall, kath-kuni, koti-banal, taaq and wooden buildings are more sustainable. Vernacular practices evolved in hilly areas like the use of local materials, thermal comfort, environmentally friendly design, smaller foot print, contextual appropriate development is considered as the essential requisites and have low environmental impact. Uses of local materials like stone, bamboo and slate with other contemporary materials are better for hill area development. Ecological considerations and eco- tourism are long-term policy measures for sustainable development. Sustainable transport system which makes it environment friendly and socially beneficial to the community must be opted for hill regions. Among other factors affordable and contextual housing in hill towns also guides sustainability. Buildings in hill regions should be designed on the principles of green architecture.

For achieving sustainable development in hill towns, it is must to have a statistically significant analysis to determine how urban image, people perception and energy efficient designs can contribute in preserving the character of hills and move towards sustainability. In total one can conclude that for sustainable strategies in hill towns there is need of better association in land use pattern, economics, energy efficient buildings, people perception, environment and sustainable public transport system.

1. URBAN FORM IN HILLS

Quality of urban form in hill towns is configuration of its visual, ecological and climatic setting. In an urban hill area along with socio-economic conditions and land use planning, its urban form takes into consideration density, street layout, traffic patterns, and transportation options. Urban form of major hill towns is not usually based upon economic location theories but rather depends upon the dynamism of hill areas, geographic resources, panoramic views, pleasant climate and their cultural value. Sustainable urban form needs to respond to a variety of existing settlement patterns and contexts.

Hill towns have unique urban form in terms of their character and aesthetic values due to their peculiar settings, unique resource assets that impart them their image and make a settlement a unique experience by itself. A sustainable city should be compact, dense, diverse, and highly integrated where one can consider city as a hill rather than city on a hill.

In order to achieve sustainable urban form for hill town its energy consumption, quality of life and ecological footprints should be considered as three main factors, which can further be divided into sub factors based upon socio economic conditions and land use planning. Socio-economic condition factors like household income, number of children, sex & age along with land use parameters like planning, population size and jobs location & housing balance plays an important role in the formulation of urban form.

It is widely recognized that combinations of elements of urban form such as infrastructure, density, land uses, urban layout, building types and transportation have an influence on the economic performance, environmental biodiversity, energy use, social life and cultural climate of a city which in turns defines sustainability. In order to achieve sustainable urban form for hill town its energy consumption, quality of life and ecological footprints should be considered as three main factors, which can further be divided into sub factors based upon socio economic conditions and land use planning. Socio-economic condition factors like household income, number of children, sex & age along with land use parameters like planning, population size and jobs location & housing balance plays an important role in the formulation of urban form.

A. Density

Density is a multi-face concept involving a number of inter-related dimensions. Density refers to the number of people (residing) in a defined area. Density needs to be an integral constituent of city planning for sustainable development. Density has a vital position to play in terms of city land use planning, public transportation and utilities provision which lead to the ecological footprint of a city.

B. Land Use

Land use is also an important tool that determines the nature of urban form. It can be defined as the total of arrangements, activities and inputs that people undertake in a certain land cover type. Land use is an important determinant of public transportation and sustainable urban form. Effective land use planning in hill towns suffers from unrealistic master plan preparation and untimely implementation.

C. Layout

Layout describes the spatial arrangement and configuration of elements at the street scale, such as grid or cul-de-sac street patterns. The layout of a neighborhood determines its accessibility and influences pedestrian movement accordingly. Access and transportation infrastructure are closely associated with sustainability. Streets which are well-connected to services and facilities and support pedestrian access are generally more frequently accessed, leading to greater concentration of multiple uses on them. In hill towns there exist no norms on how accessible facilities like primary school, local shopping center, primary health facility etc., should be to residents. There is an increasing realization that transport links are becoming almost a precursor to land development in hill cities today. Most of the new interventions in hill towns are transit oriented developments.

2. INTRINSIC DEVELOPMENT ASPECTS IN HILLS

A. Ecology and Land profile

Ecology of hill regions is very vulnerable and henceforth any development in these areas is governed by its ecology and land profile. Contour based planning is a usual character of hill towns. Difficult land profile also affects the density distribution pattern. Sometime factors like lee ward side, sun side area and fragility of land also determines the urban form of hill towns to a great extant.

B. Climatic conditions

Hill areas have usually diverse climatic conditions which can be characterized by heavy rains, cold weather, strong winds and limited sun shine days. Heavy rain intensity and catchment areas also determines the development areas. Due consideration to the geo-climatic conditions of the location contribute positively to achieve energy-efficiency in the settlement system. Leisure walking, recreational activity spaces and shopping areas are also dependent upon its climatic conditions. Areas less prone to harsh winds, continuous shade, and dampness are less developed than areas with good living conditions. Hence climate of any hill towns plays major role in developing a sustainable urban form in hill towns.

C. Culture and Economics

Hill towns are usually located areas which were not exposed to economic developments and developed their own culture due lack of integration with city. These towns usually developed as pilgrimage towns or tourist towns due to their scenic beauty. This is where function meets infrastructure and sustainability comes into picture.

Cultural and visual richness of these towns directed their architectural and settlement pattern to great extent. This unique quality eventually developed a distinct urban form pertaining to specific hill town.

Each of the above-mentioned elements is inter-related. For example, accessibility to any space within the city is dependent on its density and transport linkages, which in turn are dependent on the land use. Layouts are dependent on the land profile and in turn are dependent on density also. Ecology of an area would determine the extent of open land, recreational, commercial or other alternate uses, and residential use henceforth defining land use and density. Land use, in turn, is dependent on transport linkages and the density to be achieved.

3. TRANSPORTATION IN HILL TOWNS

While it is understood that elements of urban form are dependent on each other, role of transport is critical in achieving sustainable urban form in hill areas. In hill towns the whole functional structure of industry and commerce rest on the well-laid foundation of transport. Transport connector plays a major role in defining the urban form and finally its sustainability. Due to contour-based planning, road length is much more in hills resulting in increased energy consumption and CO_2 emissions. As mobility and connectivity is an issue in hill developments, it not only guides land use pattern but also determines the density. Major dense developments, both residential and commercial are observed along the main connectors. In most of the hill town settlement is located along main transport route in a linear pattern. Hill cities are constrained by road space due to the nature of the terrain.

Road network in hills is a product of adaptation and necessities and does not follow any fixed geometry or hierarchy system. High intensity of the terrain and steep gradients pose severe constraints on the accessibility. For achieving sustainable development public transport is the key factor. Sustainable development and sustainability indicators have always included the measurement of public transport service quality.

As per the MOUD, average walking speed was calculated as 2.3 km/hr. it means that it takes 6 minutes to walk 230m in hill areas whereas distance is 500m in plain areas. Coverage of public transport is 750m instead of 500m because people walk more on hilly areas.

Among all elements of urban form, urban mobility and transport energy usage are major areas of concern in most of these developing urban hill centers. There is a strong relationship between transport networks and urban forms. Better intra urban connectivity and accessibility are a prerequisite for sustaining compact urban form. Hence, transportation network is an important urban form characteristic at all spatial scale.

Sustainable system in hill towns should be affordable, operates efficiently, offers choice of transport mode, and supports a vibrant economy. Another aspect of transportation strategy tries to limit emissions, minimizes energy usage, limits the use of land and impacts the ecology of hill region to a minimum level. There is a tremendous scope for achieving energy economies in the transport sector in hill towns.

4. TOWNSCAPE OF HILL SETTLEMENTS

According to the planning commission of India any area above 600m in height from mean sea level or with average slope of 30 percent and above is classified as hilly. For sustainable urban form in hill towns is highly dependent upon the location of transit nodes. Integration of people, transport system, architectural and urban design strategies, environmental sensitivity, respecting hill character, and good economic benefits should act together for sustainable development.

URBAN HILL SETTLEMENTSOF HIMACHAL PRADESH

INTRODUCTION

Himachal Pradesh is one of the northern hill states of India located on the foot hills of Himalaya. It was formed on 15th April, 1948 and later on in1971it was made a full-fledged State. It shares its border with Jammu & Kashmir on North, Punjab on West, Haryana on South, Uttarakhand on South-East and has an international border with China on the Eastside.

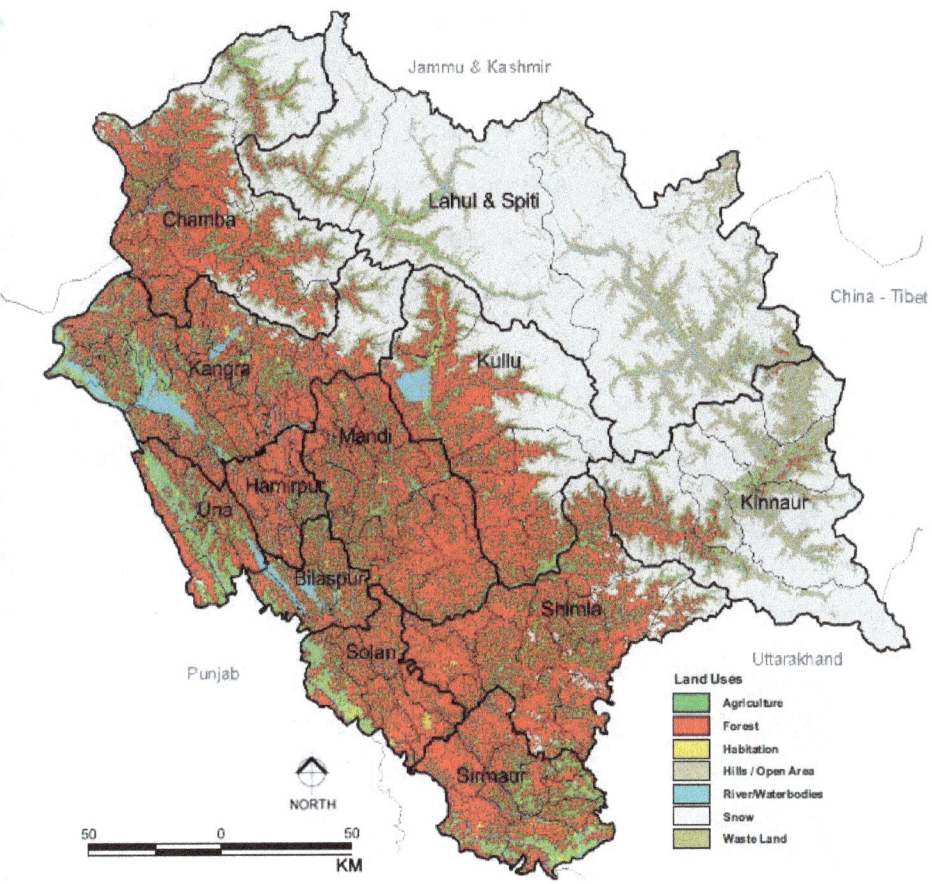

Figure 1: Map of Himachal Pradesh   Source: SEEDS, 2009

The earliest known inhabitants of the region were tribal called Dasas. In late 19th century British ruled the area after defeating Gorkhas and Rajas of the kingdoms. Himachal Pradesh is situated at Latitude 30°22' 40" N to 33° 12' 40" N and Longitude 75° 45' 55" E to 79° 04' 20" E. The mean sea level with in the region varies from 350 meter to 6975 meter. As per the 2011-Census, population of the state is 68, 64,602 persons with a density of 123 persons (per Sq. Km.). Total geographical area of the state is nearly 55,673 sq. km. Lush green forest, snow clad mountain ranges, natural lakes and springs, dry desert regions and web of rivers from Himalayas are gift of nature to this small state shown in Fig.1. Major rivers are Sutlej, Beas, Ravi, Parbati and major lakes are Renuka, Rewalsar, Khajjiar, Prashar, Mani Mahesh, Chander Tal, Gobind Sagar, and Nako. Nearly 50000 sq.kms. of state falls under protected forests area. The hill state is divided in 12 districts with Una as the smallest and Lahul&spiti as the largest district in terms of area. Shimla is the capital city and also the most urbanized area in state. Three districts namely Chamba, Kinnuar and Lahul & spiti have huge area under tribal belt. Upper area of Shimla and Sirmour district also falls under tribal belt. Shimla, kullu and kangara are the main tourist destinations and possess an urban character. Hamirpur and Una districts are small but most dense regions due to better climatic and geographical conditions.

OVERVIEW OF SHIMLA TOWN

INTRODUCTION

Shimla is situated on the last Traverse spur of the Central Himalayas, south of river Satluj at 31^0 04' North to 31^0 10' North latitude and 77^0 05' East to 77^0 15' longitude, at an altitude of 2130 m above mean sea level. Name of the town is believed to have been derived from 'Shyeemsloy', the house built of blue stones which was erected by a Faquir on the Jakhu Hill. According to another version, the word Shimla has come from "Shamla" which means "Blue Female", probably associated with the Goddess Kali. During the regime of Lord Amherst in 1827, Shimla for the first time served as seat of the Central Government. Since 1864 to 1947, Shimla was the Summer Capital of Government of India. Shimla is the only Class I town in entire State of Himachal Pradesh with majority of towns falling under Class IV category. Shimla is characterized by unique and distinct British architectural heritage zone with zoning regulations. City functions as education, administrative, heritage and tourist Center. City is a traditional educational center since British time and has an

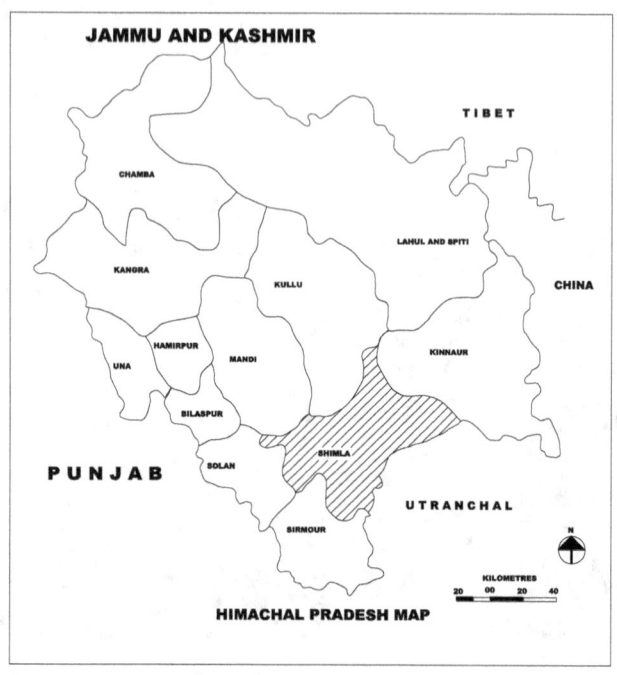

Figure 2: Location of Shimla, Himachal Pradesh

average literacy rate of 77.76%. The sex ratio of Shimla district is 916. City form is linear in shale and spreads along east- west axis along the ridge line. Most important connector, Chandigarh – Shimla-kaurick NH 22 acts as lifeline of the city. Fig. 2 shows the lacation of Shimla district. In order to ensure planned and regulated growth, government of Himachal Pradesh constituted Shimla Planning Area in 1977. Shimla Planning Area comprises of Shimla Municipal Corporation, Special Area of Dhalli, New Shimla, Tutu, Kufri, Shoghi and Ghanahatti which is spread across 100 sq. km. Shimla is known for its picturesque Mall, Scandal Point, The Ridge, Coffee Houses, Clubs and Theatres. Shimla has magnificent climate and environment which provides fascinating scenery throughout the year. Snow-clad Greater Himalayas around Shimla hills have great view and is of continuous delight for tourists. Natural vegetation is dominated by dense deodars, large oaks and elegant pine trees across the town. The city core is heavily loaded with activates and maximum city level are also faced on the same zone.

1. CLIMATIC CONDITIONS

Shimla falls under cold and climatic zone in Himalayas. During winter season Shimla gets snowfall. The average annual rainfall in the region is 900mm. Increasing heat in summers, declining quantum of snow in winters, unusual behavior of monsoon and frequent dry spells are the prime climatic concerns. Sky conditions are overcast for most part of the year except during the brief summer period. From Fig. 3 it is evident that the highest temperature during summer months of May-June goes even more than 30^0 C. Temperature, however, during winters goes down even up-to 4^0 C. Solar radiations are low

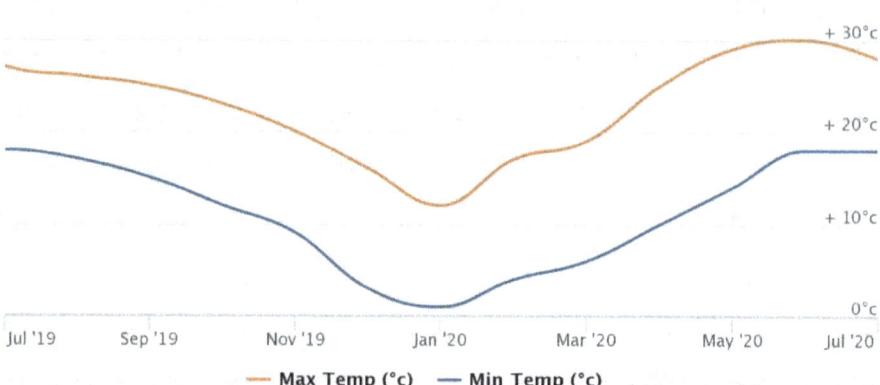

Figure 3: Average temperature in Shimla   Source: worldweatheronline.com

in winter with diffuse radiation. Relative humidity varies between 70-80%. Fig. 4 shows high rain fall in months of July and August whereas in remaining months precipitation is low. Summers are usually pleasant with high rainfall. Winds are generally intense in nature, especially during rainfall. Topography plays significance role in determining direction and speed of wind flow.

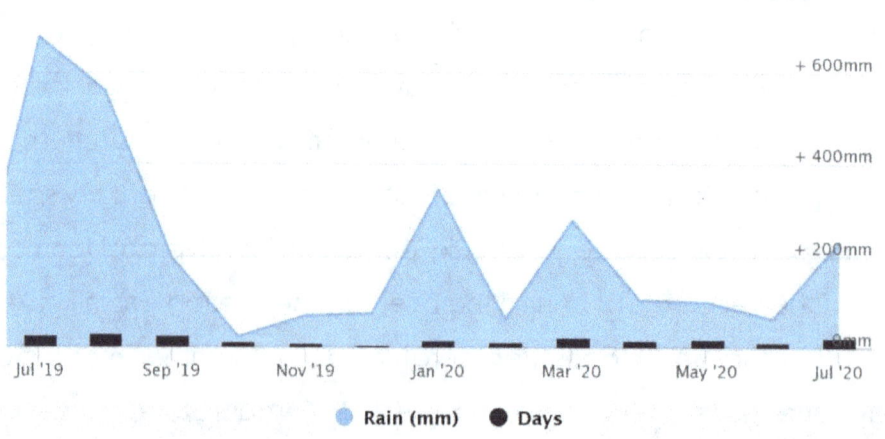

Figure 4: Average rainfall in Shimla   Source: worldweatheronline.com

2. TOPOGRAPHY

Shimla town is classified as class I hilly town. City is composed of seven small hills and main hill is having core development around it. As per TCP the 25 square kilometre of the city area is spread over seven hill spurs. These are prospect hill having cantt (Military contonment) area as main function, observatory hill having Institute of Advance Studies (Former Indian President Resedential and official complex), Jakhu hill having old temple of Lord Hanuman (Ramayan time), Summer hill having main educational hub and state university, Inverarm hill having state museum and Tele brooadcasting hub, Elysium hill major with medical hub and last one Bantony hill having main commercial, recreational and admistrative zone. World famous padestrian 'The Mall' and 'Ridge' are reffed at this peak only.Jakhoo Hill is the most elevated spur of Shimla. These spurs are inter-connected by roads. The mean elevation of main Hill spurs; Prospect Hill, Observatory Hill, Summer Hill, Potters Hill, Museum Hill, Jakhoo Hill and Elysium Hill are 2177m, 2150m, 2104m, 2073m, 2201m, 2454m and 2257m, respectivelly. Fig. 5 shows the position of various hill tops in Shimla city along with contour intervals. Most of the area is above slope of 30^0 and some areas have slope upto 60^0. Most of the populated area falls in 2000m to 2200 m range with heighest point at 2500m high (Jakhu hill top). The major land-uses are located on the southern face of Shimla due to gradual slopes and sunny side.

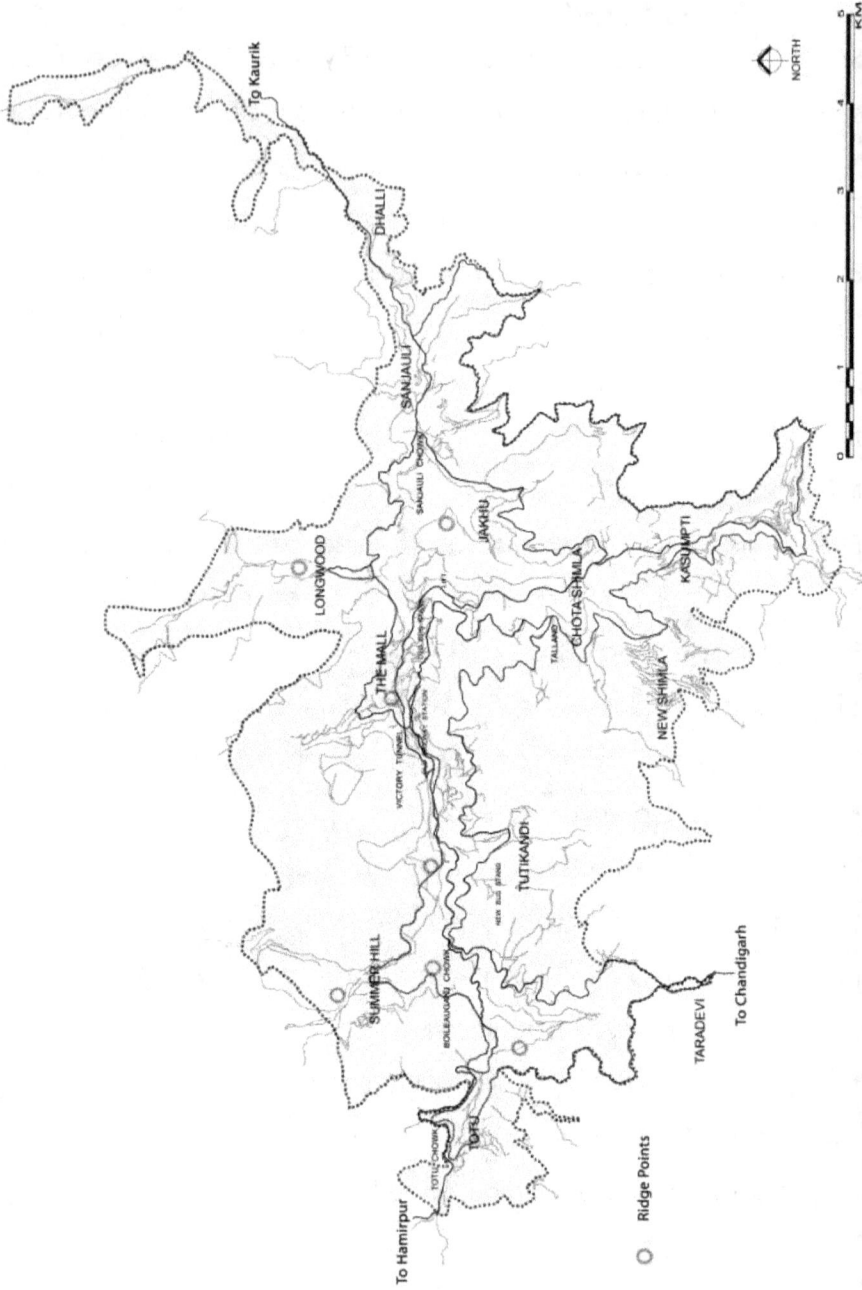

Figure 5: Contour map of Shimla city

3. TOURIST INFLOW

Shimla is one of the most popular hill station and tourist destinations in India and accounts for almost a quarter of all tourists arriving in Himachal. Immense natural and climatic diversity of Shimla have attracted both domestic and international tourists. There has been substantial growth in tourists over the last few years. Increased per capita income and tourism policies have induced this growth. There is a continuous growth in tourist inflow with maximum visitors up to 400000 in month of May shown in Fig. 6. Out of total tourists traveled to Shimla about 5% constitute foreign tourists. April to July is the peak tourist season. The average stay of tourist in Shimla is about 1.35 days. The bed capacity of all the hotels together was around 12000 rooms.

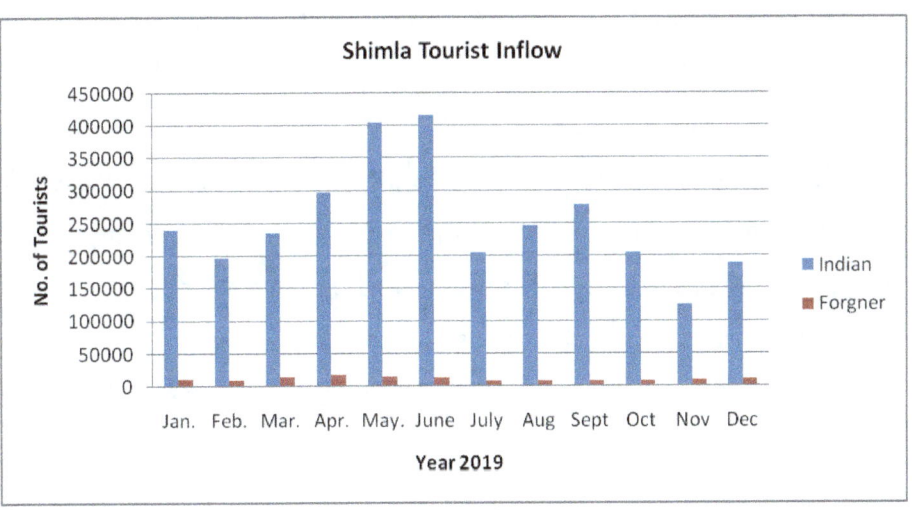

Figure 6: Tourist inflow per day   Source: Tourism department, H.P.

MORPHOLOGY AND URBAN FORM

1. LAND USE AND GREEN COVER

Green cover in the city is last surviving urban forest in the country. The urban forests contribute to value of Shimla by absorbing storm water and improving air and water quality. Of the total area of 9950 hectares of Shimla Planning Area, about 1475 hectares which accounts for 15% of the total SPA is under urban use. Out of this total area, 60% of the land is occupied by forest, 20 % is used for agriculture, 4% is used for traffic and transportation, 9% of the total land is used for residential purposes and remaining 7% is used for other activities. The land-use for housing is very high in few areas having gross residential density as high as 800 pph. Parks, gardens and open spaces in the city were covering about 6 ha. Out of total urban area 61% comes under residential and 25% for traffic and transportation. Maturity of Deodar trees, encroachment in forest, degradation of forest due to tree felling and debris along the hill slopes, soil erosion and landslides are major concerns for city forest area. Fig. 7 shows the overall land use and green cover of Shimla city and helps in identifying the routes which will help in preserving the green belt of the city. Prominent thick tree cover is present in valley near Chhota Shimla. At Mt. Jakhu there are white oaks and deodar enhancing the beauty of town scape.

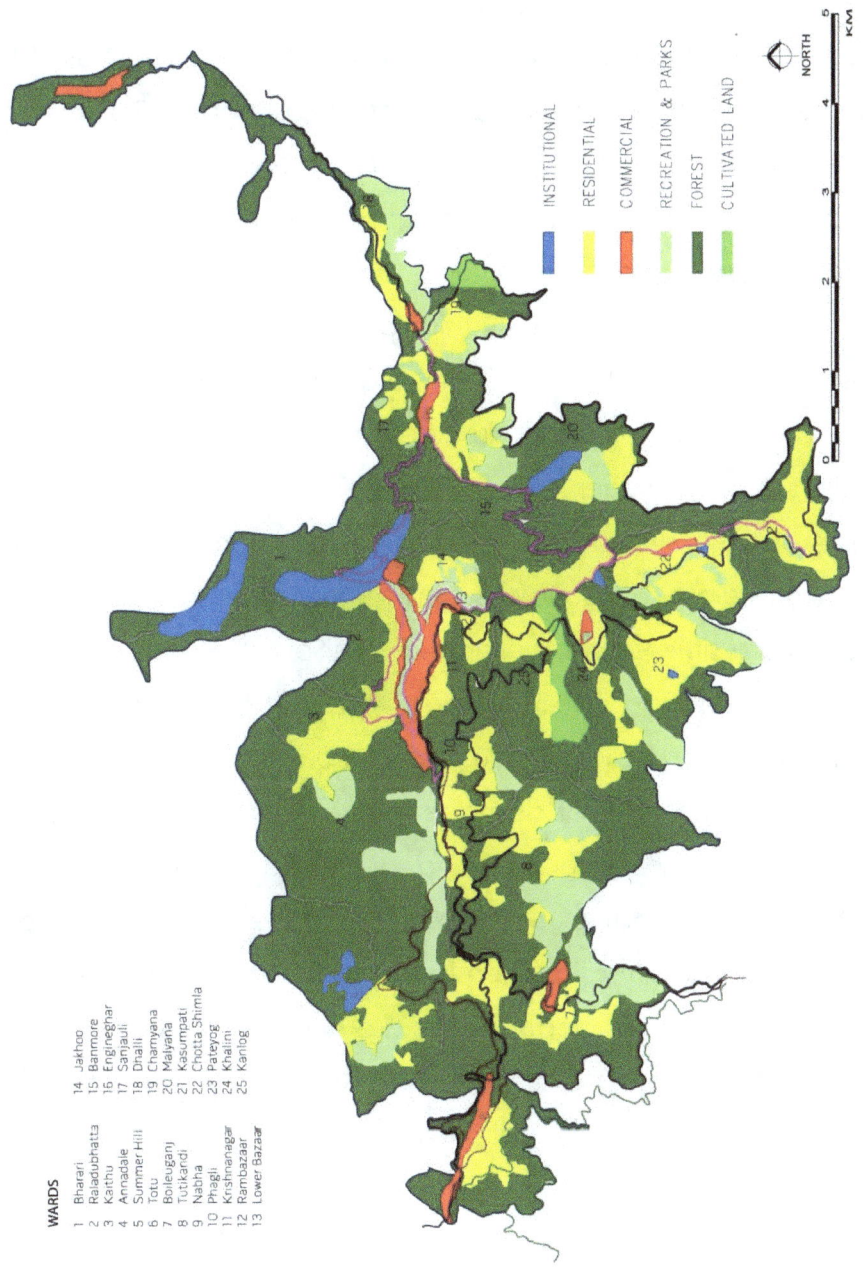

Figure 7: Land use map of Shimla city

About 42% of the total green area is under forest cover and 36% is open area occupied by shrubs, bushes and grasslands. Out of this 150 hectares open area, 124 hectares is under Government ownership and just 26 hectares is in private ownership. Thickly built up pockets amidst green areas namely Boileauganj, Tutikandi, Phagli, Lalpani, Himland locality, Dhobighat, Jakhoo, Bothwell Estate, Goodwood, Shankli, Ruldu-ka-Bhatta, ChauraMaidan and Ellesium Hill near ChauraMaidan are highly congested, facing severe infrastructural and environmental problems. The prime green pockets on the higher altitude and on slopes, though form the crown of the town, already carried constructions thereon and in their immediate surroundings. The Core wherein many green pockets are situated, comprising of most of the Shimla, possess a precious natural and built heritage, requiring preservation at any cost. Cutting and filling activities in and around green pockets have already caused a lot of damage to precious coniferous Deodar green cover. The natural setting of Shimla has already been disturbed to a great extent due to extensive deforestation and uncontrolled construction. Sensitive ecosystem is degrading with great speed and old deodar trees are diminishing due to air and land pollution. Green cover is reducing day by day and city is losing its beauty and environmental quality.

2. DEMOGRAPHIC CHARACTER

Shimla had population of around 1.7 lakh in 2011 with a density of 159 persons persq.km. The floating population of Shimla was around 3 million in 2015. The main housing areas in Shimla are the core city, part of Kaithu, Shankli, Longwood, Chhota Shimla, Jakhoo, Kasumpti, Sanjauli, Summer Hill, Boileauganj and Tutikandi. About 82% of whole population of SPA lives in M C Shimla including Dhalli, Tutu, and New Shimla. Total number of housing units in planning area is about 0.45 lakhs. Land available for development in Shimla is mainly private land. As the land supply from other sectors is very slow, the supply of land for housing from private landowners forms major share i.e more than 76%. Scarcity of land due to rigorous topography, sinking & sliding zones, and lack of water availability, higher development & construction cost, inability of HIMUDA in land management, absence of proper planning provision has led to very chaotic condition in city. Core area constituting of wards namely Nabha, Krishna Nagar, Ram Bazar have very high density and very less scope of future increase in population as shown in Fig. 8. On other hand wards namely Summerhill and Tutikandi have low density and future expansion of city is projected in these areas. Core area is presently catering various functions related to employment, education, commerce and recreation.

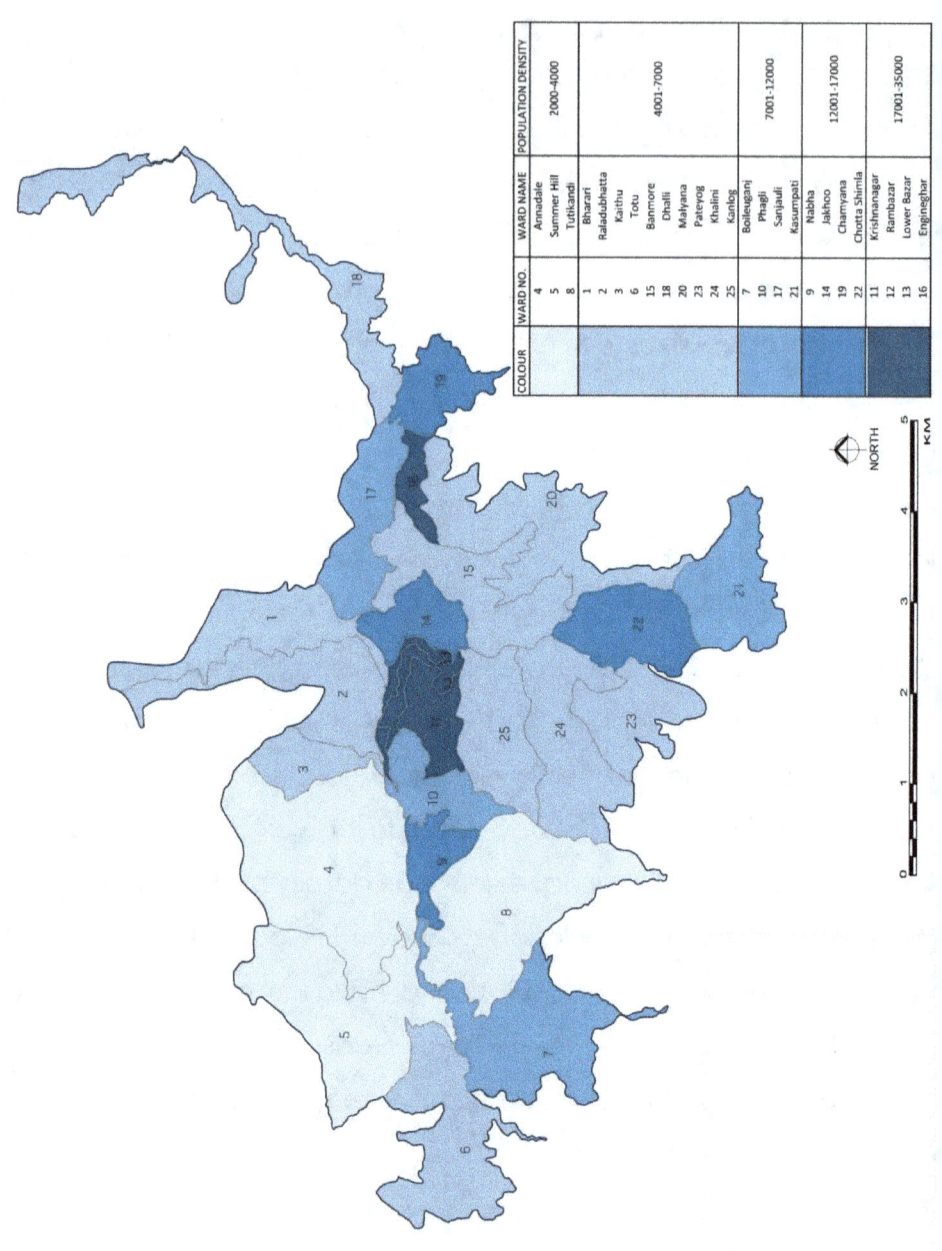

Figure 8: Demographic character of Shimla   Source: TCP department, H.P.

3. GRAIN AND TEXTURE

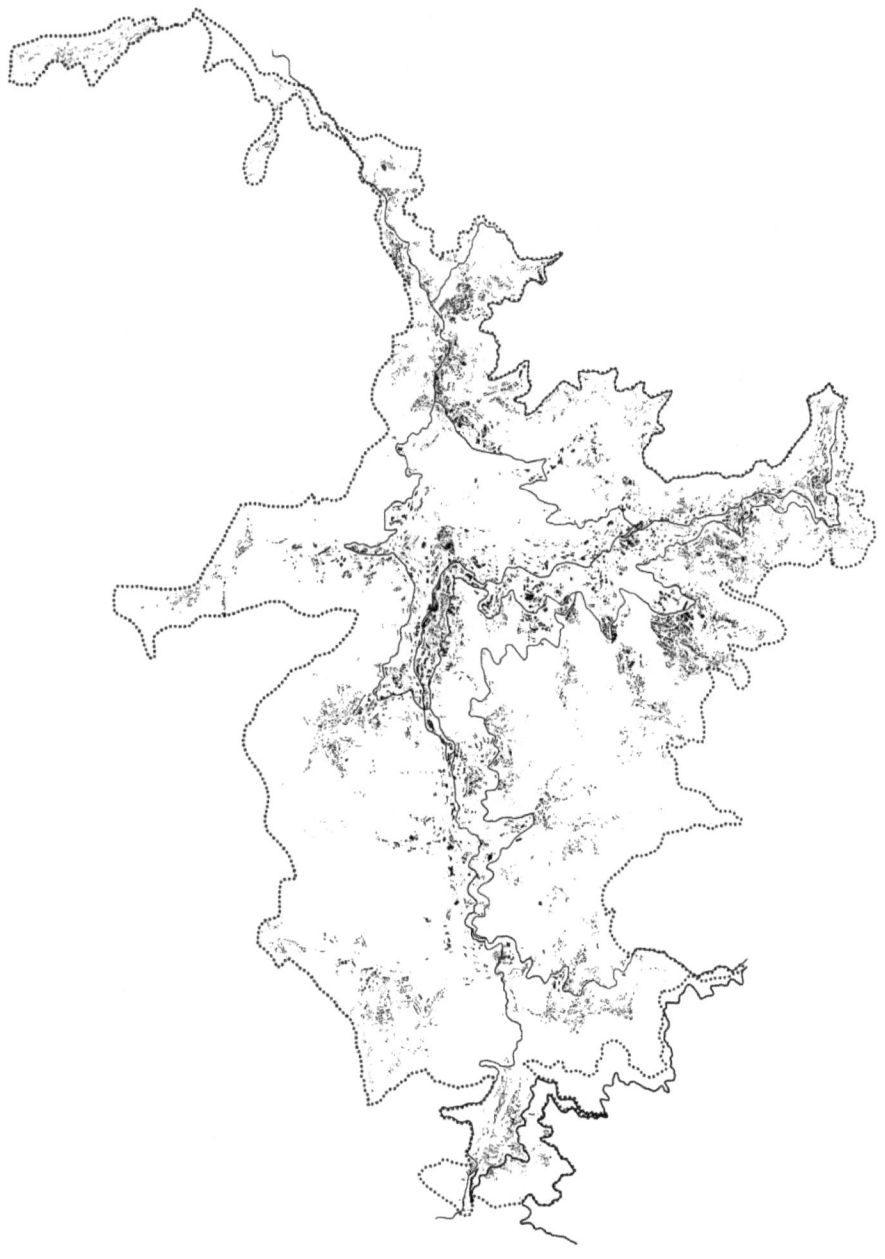

Figure 9: Figure ground map of Shimla

4. TRANSPORTATION AND MOVEMENT

Hilly terrain of Shimla allows both horizontal as well as vertical movement. Mobility in Shimla is only one of its kinds having "Fine Roads" (roads restricted for general entry), parallel connectors with limited space for vehicles to move. Horizontal movement is primarily by arterial roads and vertical mobility options are lifts (between Mall road and Cart Road) and pathways /staircases connecting various streets. The Mall road and the Ridge in Shimla is restricted to a pedestrian and can be approached either by paths (a combination of stairways and black top streets) or by the Lift. Fig.10 shows the main transport nodes and city level connectors. With major roads functioning more than their capacity, on-street parking causing congestion, walking being the predominant mode of travel and increasing floating population demand major interventions in transportation system of city. Shimla town is facing traffic congestion due to its peculiar geography and old road infrastructure. In order to cater the needs of growing population very little progress was made in subsequent years towards road development in central city. Development of main activity areas along arterial roads has increased traffic congestion which is characterized by regular traffic jams.

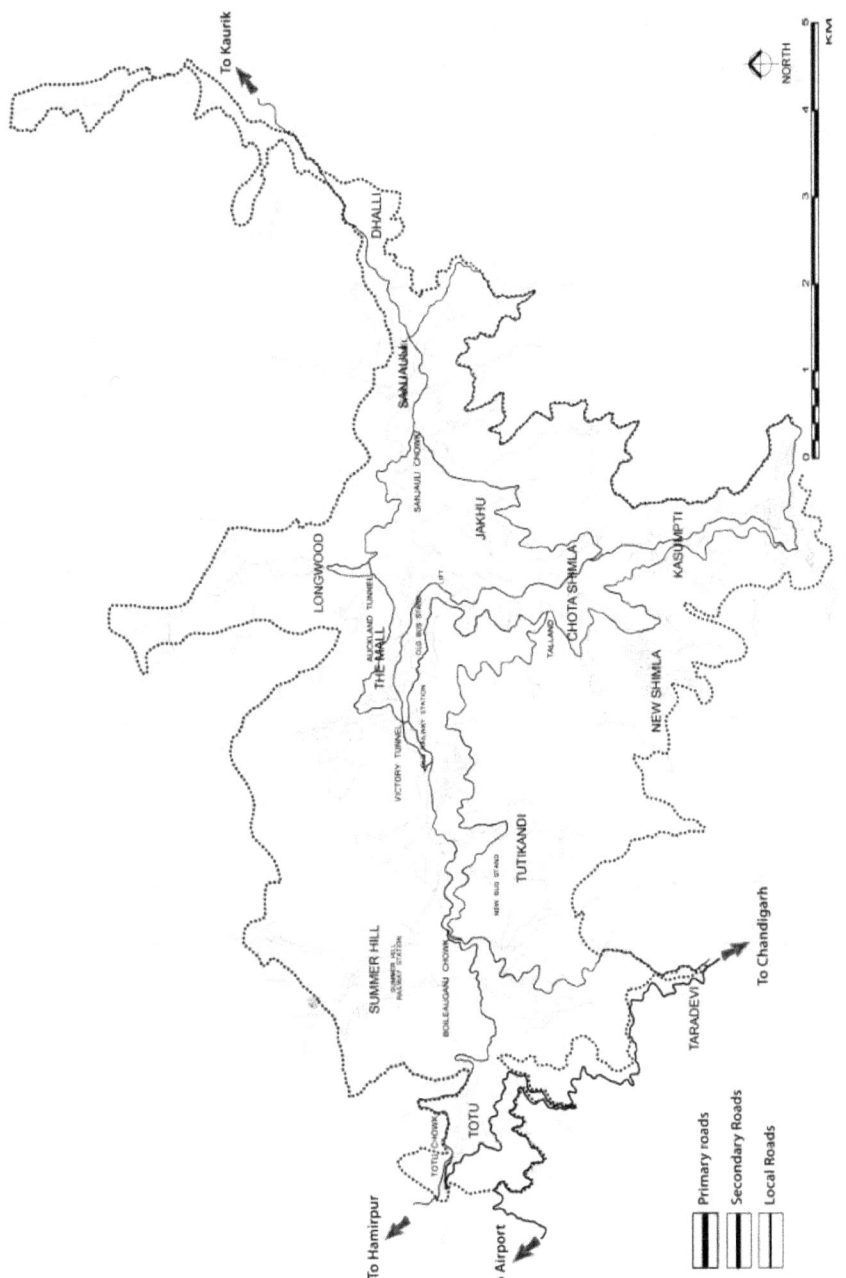

Figure 10: Main transit nodes in city

Most of trips are pedestrian in nature with trip length less than 3 km on an average. At present population of the town is nearly 1.7lac. Survey conducted by Town & Country Planning Department indicates, 14 % persons use buses as mode of transport and 58% move in Shimla Town on foot, rest 15% use their own private vehicles for movement. The number of registered vehicles in Shimla Planning Area in the last decade has increased at decadal growth rate of 34%. With the present traffic scenario city faces lot of traffic problems due to lack of space and increased population. All the forms of work trips are not accessible by vehicles because of the restricted roads. Therefore, all forms of work trips include walking as the important mode of work trip. With the present traffic scenario city faces a lot of traffic problems due to lack of space and increased population. Public Transport in Shimla Planning Area is only by the bus transport by HRTC and mini buses operated on selected routes by the licensed private operators. 62%of the work trip is by Public Transport catered by 27 % of the traffic volume (bus trips) within Shimla City as against 30% of the work trips are by private four wheelers, which form 65% of the traffic volume. Most of the off-street parking lots are working to full capacity and On-street parking can be witnessed all along the roads in Shimla.

CITYSCAPE

1. LANDMARK AND ACTIVITY GENERATORS

To promote and maintain tourism in the city government has established Tourism Development Council for Shimla. Fig. 11 shows the places of tourist attraction in Shimla which include The Ridge, The Mall, Jakhoo Temple, Sankat Mochan Temple, Tara Devi temple, Kufri Ski Slopes, Shimla state Museum, Prospect Hill and Chadwick Falls. The Mall and the Ridge are main tourist zones and various governmental offices are situated in the area.

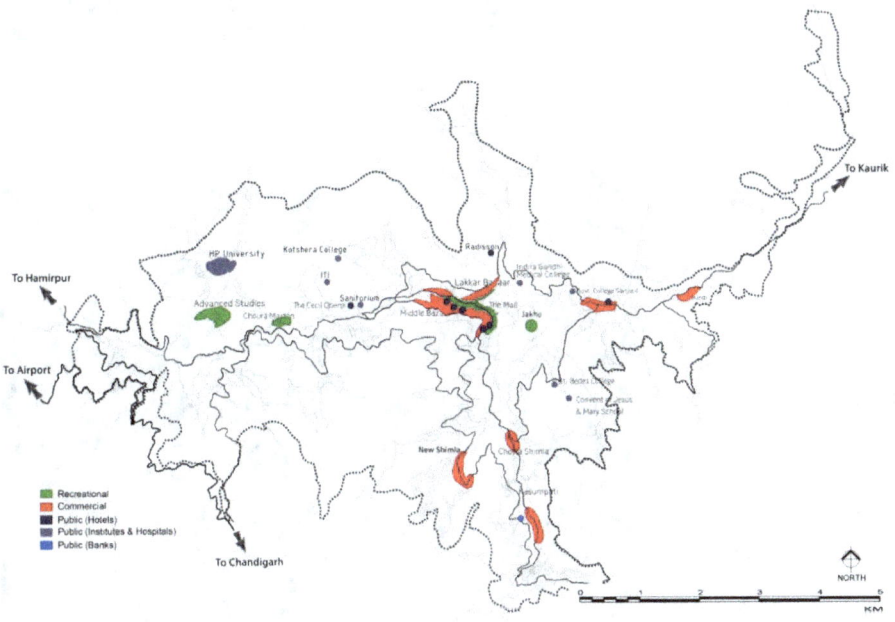

Figure 11: Major activity zones in Shimla

2. VERNACULAR ARCHITECTURAL VOCABULARY

Old buildings of British time have distinct architectural character. These buildings are fine piece of architecture and add richness to the city profile. The Town Hall has unique façade and a saga of British Heritage shown in Fig. 12. It is the best example of imperial planning of Central city. The Railway Board Building has distinct style with overwhelming use of iron pipes in its façade. The Bantony is a rare demonstration of hill town roof scape along with attractive windows and doors. The state library building on the Ridge is a rare architectural feast with mix architectural vocabulary.

Figure 13: Town hall Shimla

Figure 12: Building with old residential character

3. CONTEMPORARY ARCHITECTURAL EXPRESSION

Hhigh demand in housing sector and uncontrolled growth has taken away the visual character of Shimla. New and fast construction method and lack of knowledge towards old architectural vocabulary shown in Fig. 13. The new residential and public buildings have degraded the city skyline. New houses and public buildings are halfway attempts to replicate old architectural style. Use of conical roof, gable end roof, bay window in various combinations is a producing terrible modern architecture. Fig. 14 & Fig. 15 shows the newly constructed houses in the city. Form the profile of roof it is clearly evident that it has no connection with the lower part and slopes in roof are not dealt with sensitivity. Mixture of various elements in these buildings belongs neither to old style nor fall in category of modern architecture.

Figure 15: Sloping roof form in new construction

Figure 14: Framed construction for multistory

CONCLUSION

Diverse geographical and topographical settings like steep slopes, forest areas, extended hilly spurs, limited vehicular connectivity and climatic constrains have guided the growth measures in Shimla. Furthermore, extremely limited constructible land and dependence on pedestrian movement limits expansion of town to a concentrated area. As per SCDP Shimla is facing the problems typically faced by any hilly region like scarcity of buildable land, emergence of linear urban corridors, limited accessibility and sprawl of urban structure due to rapid growth in central and peripheral areas.

Shimla Development Plan 2021 has been prepared keeping in view various aspects like development of various nodes on the periphery of city, development of commercial activity centers, and relocation of non-conforming activities like timber market, transport hub, wholesale grain market, wholesale vegetable market and bus stand. Along with other interventions shifting of Government offices & institutions in heritage buildings and using them for tourism activities is also on agenda. Some of the major traffic problems that persist in the town are lack of connectivity between old and new bus stands, overcrowded buses and frequent traffic jams, limited public transportation, lack of pedestrian path on road side and limited parking lots. Among above listed complexities, city development plan has tried to resolve the issues by introducing more buses, elevator and parking lots. The high increase of population and huge inflow of floating population is exerting immense load on the existing infrastructure. This alarming situation has brought breakdown in transportation system and needs early interventions before complete collapse.

It is evident that pedestrian mobility in the city needs paths/stairs for vertical movement and footpaths for horizontal movement. Pedestrian safety is must at pedestrian crossings and over bridges for better moment and connectivity. Maximum active zone of city is the core area around old transit terminal. Lack of proper physical and social infrastructure is leading to encroachments, unauthorized constructions incompatible with the culture and context. Environmental, cultural and heritage image of city has been lost due to poor enforcement of zoning rules and regulations.

OVERVIEW OF MANALI TOWN

INTRODUCTION

Manali is a small town in Kullu District of Himachal Pradesh situated on the right bank of River Beas shown in Fig. 16. Manali is named after the Sanatan Hindu lawgiver Manu. The name Manali is regarded as the derivative of 'Manu-Alaya' which literally means 'the abode of Manu'. Legend has it that sage Manu stepped off his ark

Figure 16: Location map of Kullu, Himachal Pradesh

in Manali to recreate human life after a great flood had deluged the world. Manali lies in the North of Kullu Valley. The valley is often referred to as the 'Valley of the Gods'. Old Manali village has an ancient temple dedicated to sage Manu. The British introduced apple trees in the area. To this day, apple— along with plum and pear— remain the best source of income for the majority of inhabitants.

1. CLIMATIC CONDITIONS

Manali falls under cold and climatic zone in Himalayas. The climate in Manali is warm and temperate. In winter, there is much less rainfall in Manali than in summer. This climate is considered to be Cwa according to the Köppen-Geiger climate classification. During winter season (mid-October till mid-March) city gets snowfall. The temperatures range from 4° C to 26° C over the year. The average temperature during summer is between 10° C and 26° C and between 15° C and 12° C in the winter shown in Fig. 18.

On average, some 45 mm of precipitation is received during winter and spring months, increasing to some 115 mm in summer as the monsoon approaches. Summers are usually pleasant with high rainfall. Monthly precipitation varies between 31 mm in November and 217 mm in July. The average total annual precipitation is 1,363 mm shown in Fig. 17.

The climate of Manali is termed as 'climate of recreation' as it is comfortable and pleasing in summer. At altitude more than 1800 metres, winters are not that tolerable as that of the rest of the year. Manali experiences more than one metre snowfall during winter.

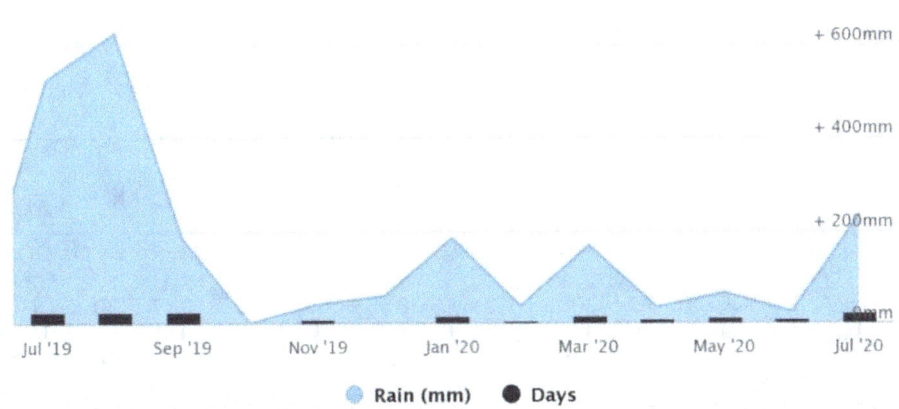

Figure 17: Average rainfall in Manali Source: worldweatheronline.com

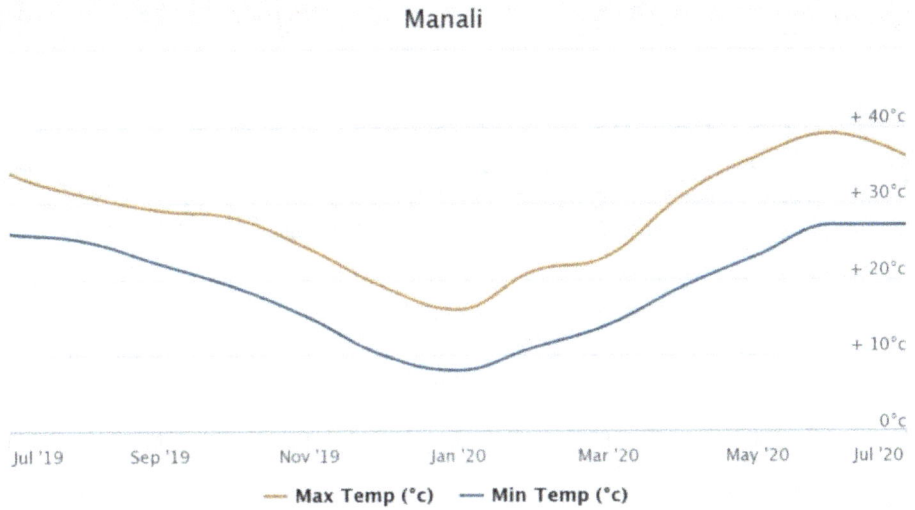

Figure 18: Average temperature in Manali Source: worldweatheronline.com

2. TOPOGRAPHY

The Manali lies on 1949m above sea level. As per the planning commission of India classification, Manali town is classified as hilly town. Manali is about 6500 feet above mean sea level. This town is endowed with natural scenic beauty. The contour map of Manali town is shown in Fig. 19. The ridge points are demarcated by major religious structures. Temple of Mata Hidimba is one of the most famous tourist points with in core area. City is divided by river at one end and slopes are towards the water stream. The mall road is only less steep portion in the city profile. Due to its undulated slopes Manali has developed along the major road connector only.

Manali and its surroundings are located on gradually sloping terraced valleys parallel to river Beas carved into spurs by cross drains and nullahs, mostly reclaimed by removing forests. The soil on the spurs is alluvial and favorable for agriculture and horticulture whereas the river-side slopes are mostly studded with large and medium sized boulders in loose and vulnerable soil strata. These conditions reduce the bearing capacity of soil and is not favorable for heavy structures. The area falls within seismic zone IV near a fault line and is subject to earthquakes.

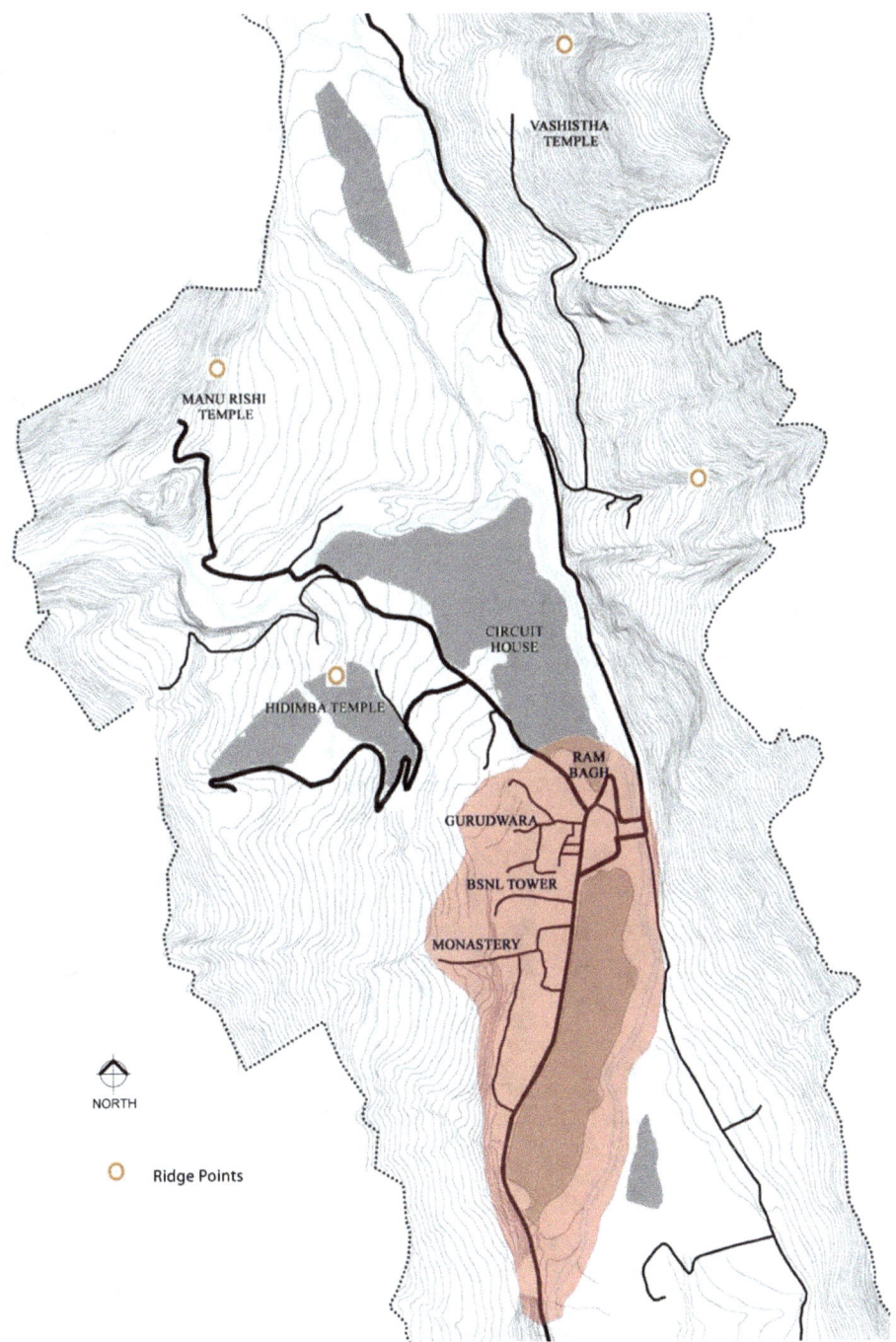

Figure 17: Contour map of Manali, Himachal Pradesh

3. TOURIST INFLOW ANALYSIS

The city has been attracting lacs of tourist from all over India and abroad in summer as well as Winter Seasons. In the recent years, it has emerged as an alternative destination to Jammu and Kashmir. According to abroad estimate about 10 to 15 lacs tourists visit Manali every years. Tourist infrastructure is fairly developed in Manali as it has 500 registered hotels having 15000 bed capacities. There are several places around Manali which are famous from tourist point of view, such as Rohtang Pass, Marhi and Gulaba fall. In the late 1980s, Manali witnessed a surge in tourist traffic. This once quiet village was transformed into a bustling town with numerous home stays as well as the occasional boutique hotel. About 18,000 tourists visit Manali in a day in peak season with May and June having maximum tourists shown in Fig. 20

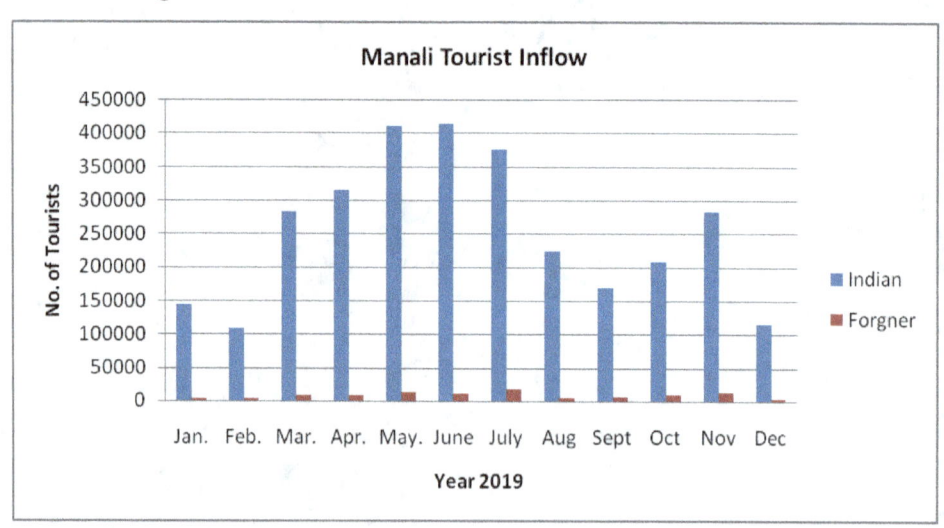

Figure 18: Tourist inflow per day   Source: Tourism department, H.P

MORPHOLOGY AND URBAN FORM

1. LAND USE AND GREEN COVER

Manali Agglomeration is carved out of the Kullu Valley. It includes Old Manali, Bahang, Suinsa, Ranghri, Nagar, Vashisht, Chadhiari, Chachoga, Aleo, Prini, Shuru, Bahanu, Sial, and Chhial. Total area covered under the Manali Agglomeration is 1152 hectares. Manali Agglomeration has predominance of tourism activities. Its characteristic, trade & commerce is centre for the supporting population and for certain higher order facilities like education and health. Landuse pattern of town is shown in Fig. 21. It also acts as a base station for Lahaul and Spiti valley. In Manali Aglomeration the old settlements like, Manali, Vashisht, Chachoga, Chadhian, Aleo, Prini, Shuru, Sunisa, Chhial, Nasogi, Sial and Balsan are old villages with high intensity residential areas whereas in Nagar Panchayat area all uses are mixed up in a mixed landuse pattern. Maximum number of hotels and commercial establishments are concentrated within Nagar Panchayat area without any segregation for landuse. Areas around Manali have most majestic Cedars. The thick Deodar forests around the main town of Manali have made it most attractive. There is a rich nursery of indigenous as well as exotic plants within the northern forest. The forests are well protected and preserved. In the surrounding areas there are rich orchards of apple, apricot, plum and walnut.

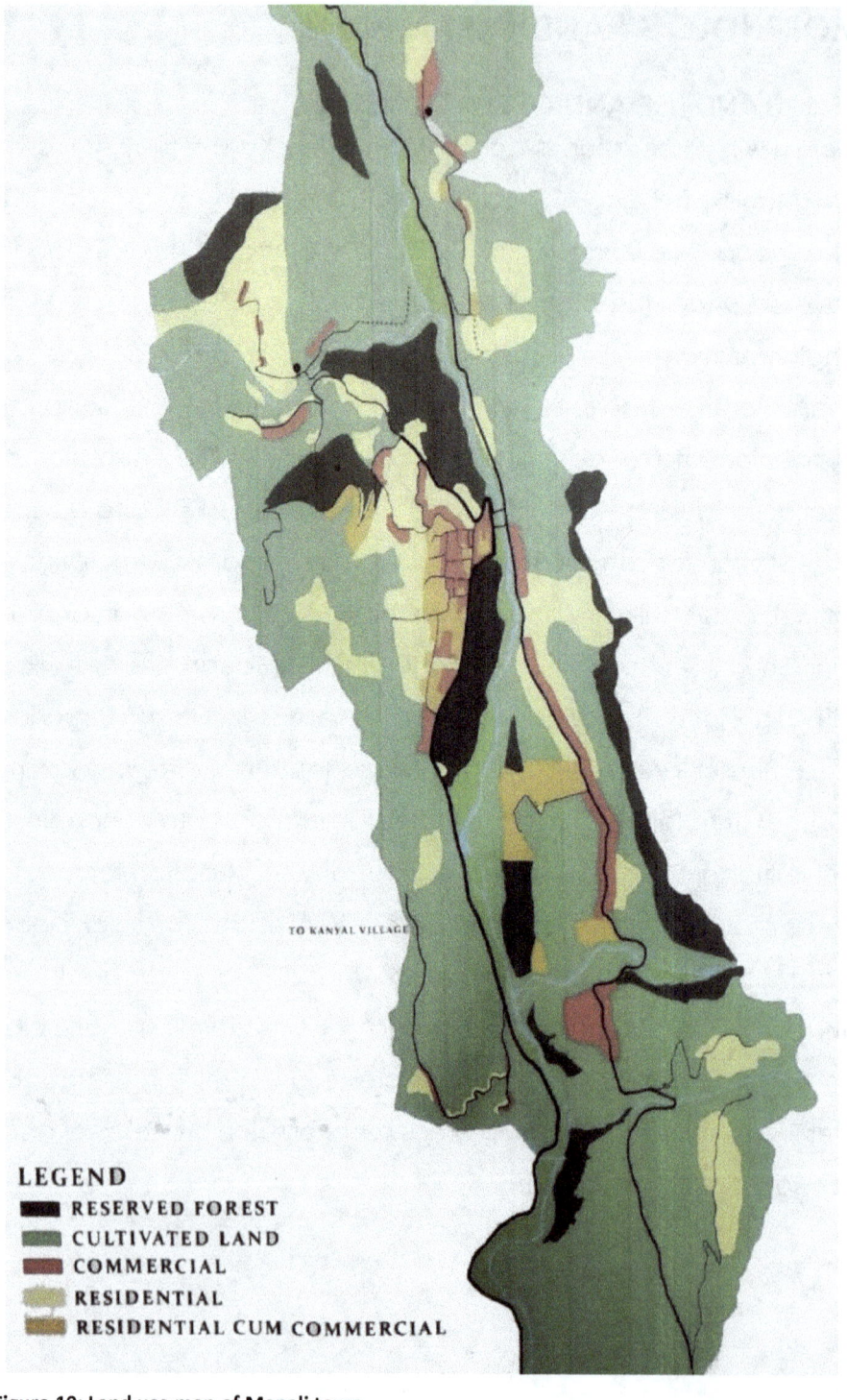

Figure 19: Land use map of Manali town

2. DEMOGRAPHIC CHARACTER

Manali has grown from a trading outpost/ village to a small town; as of the 2011 census of India, its population was 8,096. In 2001, Manali had an official population of 6,265. Males constituted 64% of the population and females 36%. Manali had an average literacy rate of 74%, higher than the national average of 59.5%; male literacy was 80%, and female literacy was 63.9%. 9.5% of the population was under six years of age. During the summer months there is a marked surge in the transients as many of them are employed in the hospitality businesses. Maximum residences are in ward 2 and ward 7 due to availability of better infrastructure as evedent in Fig.22.

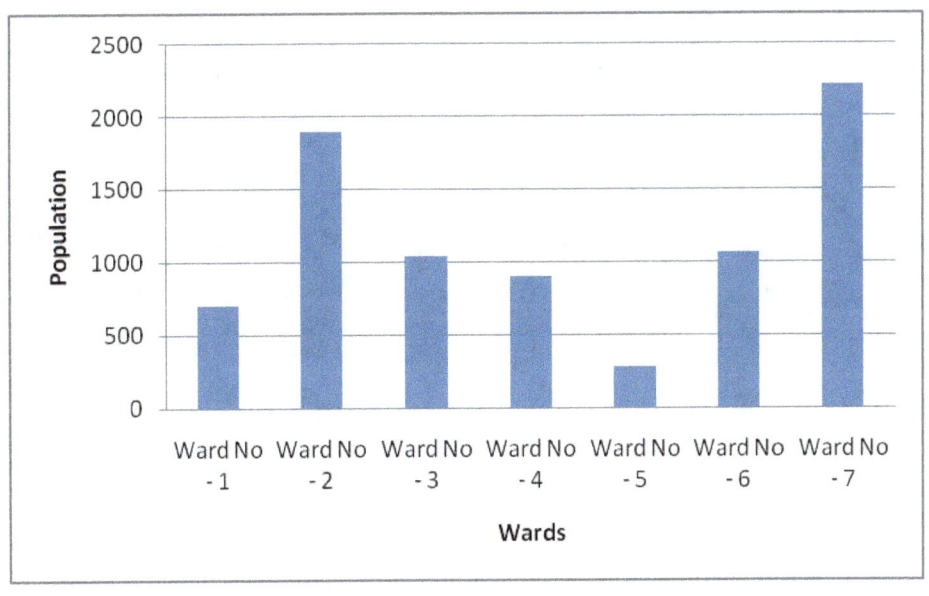

Figure 20: Demographic character of Manali town   Source: TCP department, H.P

3. GRAIN AND TEXTURE

Figure 21: Figure ground map of Manali town

Manali has a distinctive characteristic structural difference in old and new housing. The villages which are an integral part of Manali have traditionally constructed wooden houses with two storeys bearing worn out look. These houses lack modern facilities whereas the new constructions are carried out in new materials and are generally 4 to 5 storeys high. Almost all new houses are either hotels or high-class guest houses. The old residential houses are gradually being converted into commercial houses. Along the main street (Mall Road) buildings are densely packed with coarse grain, Fig. 23.

4. TRANSPORTATION AND MOVEMENT

The National Highway No. 21 is the main life line for the entire area along the right bank of river Beas. As the work areas are nearby residences, majority of workers go on foot. Whereas, 82.58% workers go on foot to their work areas, only 9.04% are using 2 wheelers and 3.72% are going to their work by Bus, Cycles and 4-Wheelers as mode of traveling to work areas account for 2.26% and 2.40% respectively. The 87.74% of the families do not have any vehicle whereas 8.25% have Cars, 2.32% use cycles and 1.69% have no other means of conveyance. Fig. 24 shows the structure of Manali town.

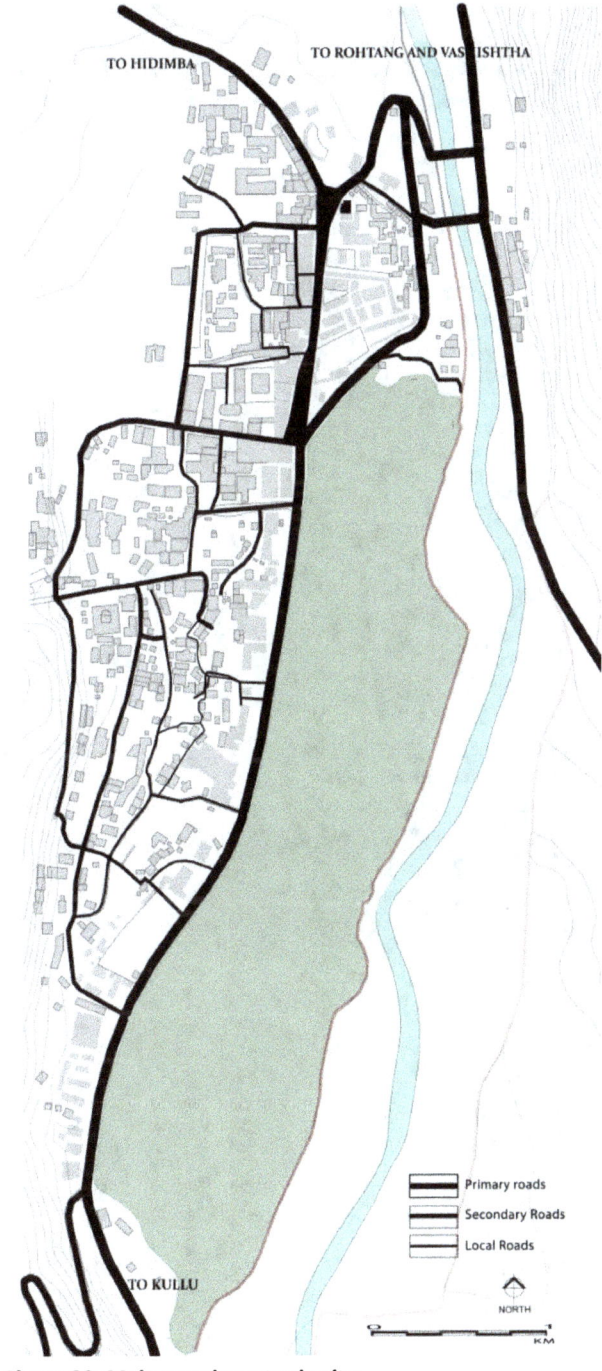

Figure 22: Main transit routes in city

CITYSCAPE

1. LANDMARK AND ACTIVITY GENERATORS

Figure 23: Major activity zones in Manali

Boiling hot water springs of Vashisht are famous attraction of the valley. Landmarks are missing along the main spines. Hidima Devi Temple and Tibetian Monastry are the main landmarks of the town. The forests around are full of wild life. Animals like black bear, leopard, wild sheep, ghoral, barking deer and birds like monal and fowls can be seen in the forests. The commercial intensity is very high in the central Manali town. Out of the total area 31% is under the commercial use. The main commercial area of the town includes mainly the shops related to tourist activities. All major activities are around main spine shown in Fig. 25.

2. VERNACULAR ARCHITECTURAL VOCABULARY

The traditional architecture has almost vanished and a modern architecture imported from the plains has replaced it. Temples acts as a major activity generator throughout the spine. Combination of buildings of different Era of timeline are exposed to different level of maintenance. Uncontrolled development and violations of building bye laws has depleted the old architectural character of Manali town. One can observe a nice blend of traditional and modern construction in the outer areas of town shown in Fig. 26. Fug. 27 shows a typical building with old residential character dominated by use of local stone and wood. This type of construction is loosing its charm due to high maintenance and lack of material availability.

Figure 24: Blend of traditional and modern

Figure 25: Building with old residential character

3. CONTEMPORARY ARCHITECTURAL EXPRESSION

Most of the structures are built recently for the commercial purpose and hence are in good condition. Only few old structures are in bad condition. The buildings constructed with Reinforced Cement Concrete, bricks and imported stones in column, beam design is dominating the scene. Many of the buildings have even flat roofs in place of conical sloping roofs conforming to hill architectural style. The settlement pattern which used to be compact and confined only to developable places has now started scattering. Fig. 28 shows the dominance of framed construction for residences in present days.

Figure 26: Framed construction for residences

CONCLUSION

The ancient landmarks of Manu and Vashisht are witnessing an unprecedented pressure on their characteristic charm. The haphazard slum like development around these historical monuments have adversely affected their aesthetic grandeur. In the by-lanes of these monuments, foreign tourists stay in the private houses, for a longer duration even for months and staying as paying guests, they are polluting the local environment. People are also getting prone to hippism and drug abuse.

Manali as a hill resort has a special character where the residential population is too less in comparison to the floating population of tourists. The physical development which took place in past 5 decades has been of spontaneous character. Unthoughtful use of land for commercial purposes without any consideration for smooth circulation and leaving open spaces for parking as well as other ancillary activities has made the situation critically worse. The encroachments on streets have narrowed the lanes. With the increase in number of vehicles entering the town the traffic congestion has become a problem especially during tourist season. In absence of private parking lots within the premises of hotels and other residential/commercial buildings the vehicles keep parked on roads which aggravates the problem. The Manali Agglomeration which is divided geographically into two parts by river Beas needs to be integrated in the interest of planned development. The river has a tendency to widen its course every year during heavy monsoons causing massive erosion of banks. The floods are quite unprecedented and the force of torrents is not predictable. There is only one interlink between the two banks through a belly bridge.

Seismically area lies in the great Alpine- Himalayan seismic belt. Region is prone to various natural hazards like earthquakes, landslides, flash floods, storms and dam failures. The hazard which however, poses biggest threat is the land slide and cloud burst. Systematic data collection and analysis will help in framing policy decisions to reduce disaster risks and build resilience. The development has to come up spontaneously as per the public demand and availability of developable land. Therefore, a mixed residential land use pattern in the new areas has been envisaged in the Development Plan.

OVERVIEW OF DHARMASHALA TOWN

INTRODUCTION

Dharamshala is the district headquarters of Kangra district. Fig. 29 shows that it is on the western side of the state and connects it mainly with Punjab. It was formerly known as Bhagsu. Dharamsala is located in the

Figure 27: Location map of Kangra, Himachal Pradesh

shadow of the Dhauladhar mountains. Dharamshala has been declared as the second capital of Himachal Pradesh state. It is surrounded by dense coniferous forest consisting mainly of stately Deodar cedar trees. This place is also famous for its newly built cricket stadium which offers opportunities to the youth of state to prepare for their future in the game. The village of McLeod Ganj, lying in the upper reaches, is known worldwide for the presence of the Dalai Lama. Dalai Lama presence and the Tibetan population have made Dharamshala a popular destination for Indian and foreign tourists, including students studying Tibet.

1. CLIMATIC CONDITIONS

Autumn temperatures average around 16–17 °C (61–63 °F). Winter starts in December and continues until late February. Snow and sleet are common during the winter in upper Dharamshala (including McLeodganj, Bhagsu Nag and Naddi). Lower Dharamshala receives little frozen precipitation except hail. The snowfall in January is usually the heaviest. Fig. 31 shows the rainfall senario of kangra which is almost same for Dharmashala town. It was caused by deep low pressure entering the Kangra district. Winter is followed by a short, pleasant spring until April. Historically, the Dhauladhar mountains used to remain snow-covered all year long; however, in recent years they have been losing their snow blanket during dry spells.The best times to visit are the autumn and spring months. Dharamshala has a monsoon-influenced, humid subtropical climate (Cwa). Summer starts in early April, peaks in early June when temperatures can reach 36 °C, and lasts till mid-June, Fig.30. From July to mid-September is the monsoon season, when up to 3,000 mm of rainfall can be experienced, making Dharamshala one of the wettest places in the state. the Kangra district gets the largest precipitation of 250.90 mm in the month of July. In November, the monthly average is lowest, i.e., 7.50 mm. Autumn is mild and lasts from October to the end of November. Even within the districts, owing to its hilly terrain, there are many variations in altitude, and the variation in rainfall is enormous.

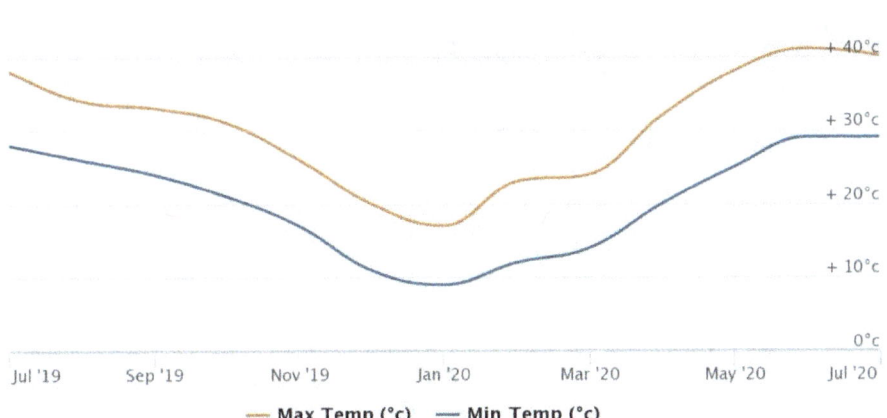

Figure 28: Yearly Average temperature range   Source: worldweatheronline.com

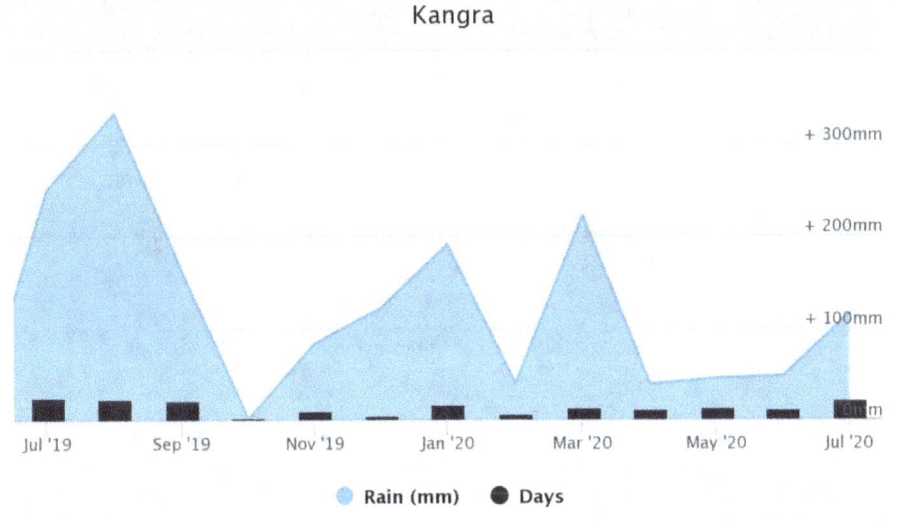

Figure 29: Yearly Average rainfall   Source: worldweatheronline.com

2. TOPOGRAPHY

According to the planning commission of India any area above 600m in height from mean sea level or with average slope of 30 percent and above is classified as hilly. Town is covering an area of almost 8.51 km².

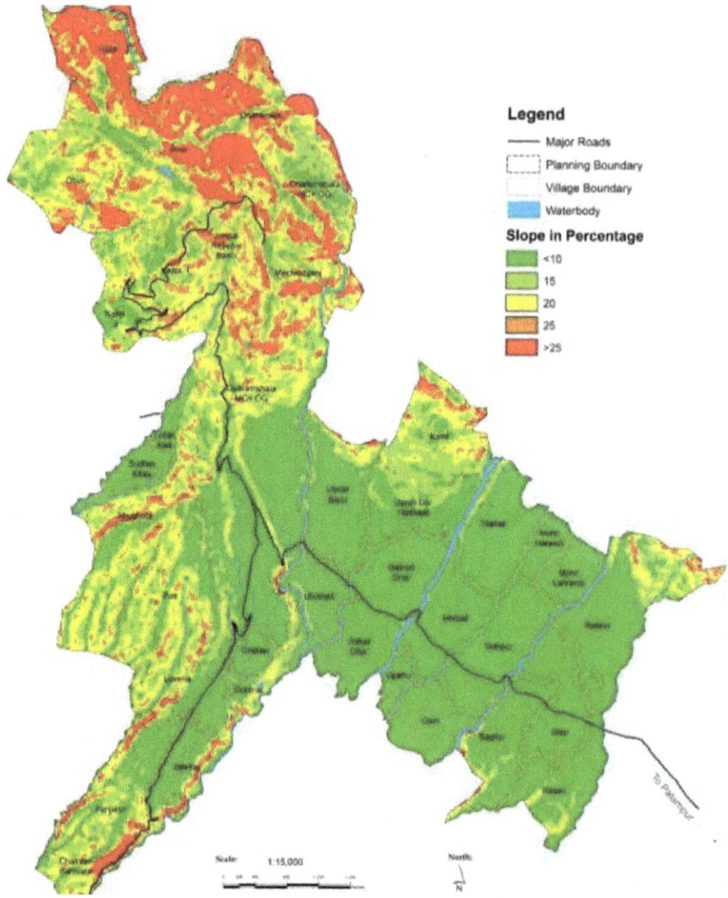

Figure 30: Contour map of Dharamshala   Source: TCP DP 2035

Geographically Dharamshala is positioned 23°13' North latitude and 76°19' East longitude. The slope of the town varies from area to area. The average elevation of the town is 1457m from mean sea level, but it ranges from 1250m to 2082m. Dharamshala has various streams and waterfalls that come down from glaciers from nearby areas. Soil available in the area is fertile which is suitable for rice, wheat and tea. Area is surrounded by Dhauladhar hills which are always snowcapped. Fig. 32 shows the slope of main town area.

3. TOURIST INFLOW

As per the tourist arrival data approximately 25 lakh3 tourists visited Kangra district in a year. During peak season in the month of April approximately 3, 08,001 tourists visited the district and about 10,267 tourists per day. The suburbs include McLeodganj, Bhagsunag, Dharamkot, Naddi, Forsythganj, Kotwali Bazar, Kachehri Adda, Dari, Ramnagar, Sidhpur and Sidhbari etc.

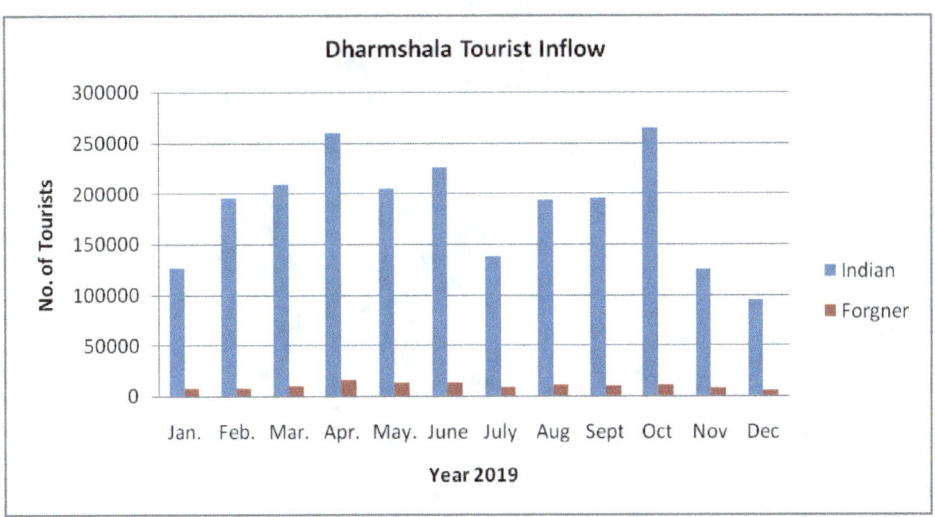

Figure 31: Tourist inflow per month   Source: Tourism department, H.P.

Dharamshala city is divided into two distinct sections. Kotwali Bazaar and the surrounding markets are referred to as "Lower Dharamshala" or just "Dharamshala." Further up the mountain is McLeod Ganj. A steep, narrow road connects McLeod Ganj from Dharamshala and is only accessible to taxis and small cars, while a longer road winds around the valley for use by buses and trucks. Fig. 33 shows that maximum tourists visit during summer season.

MORPHOLOGY AND URBAN FORM

1. LAND USE AND GREEN COVER

Dharamshala is comprising of the Planning area as well as the Municipal area. The area that comes under planning is 41.63 km², and the Municipal council constitutes 27.60 km². McLeod Ganj is surrounded by pine, Himalayan oak, and rhododendron. Markets in the town are highly congested specially Kotwali Bazar due to dense development, and narrow streets.

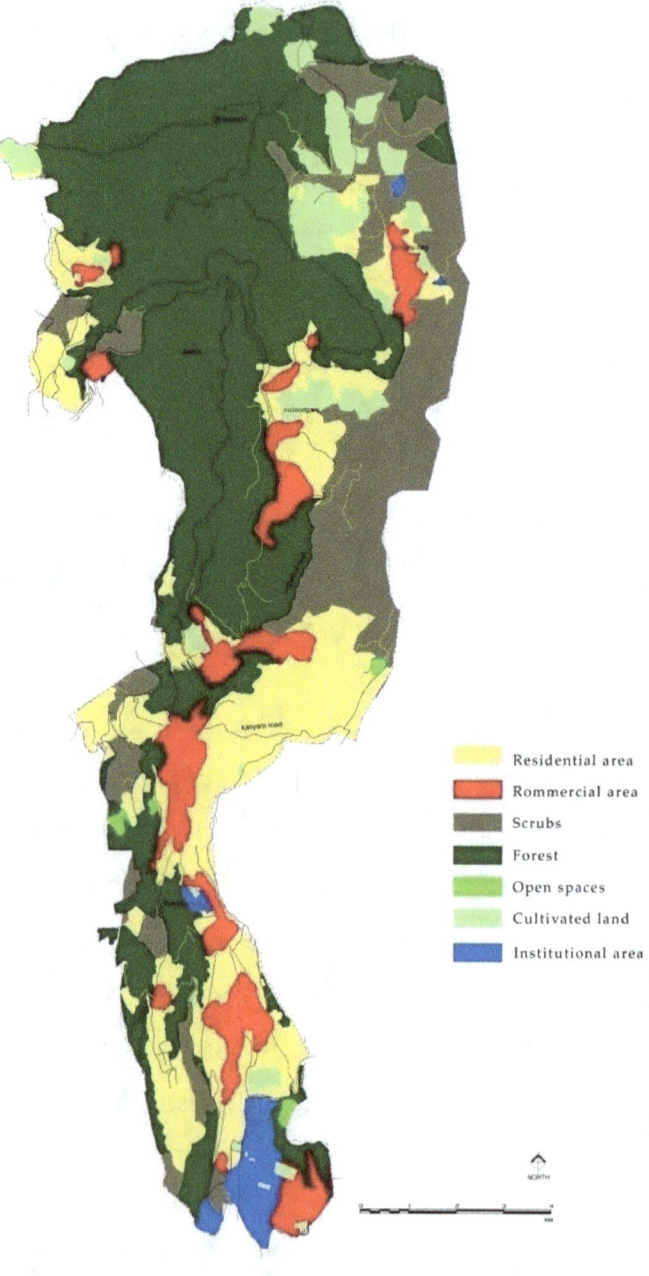

Figure 32: Land use map of Dharamshala town

2. DEMOGRAPHIC CHARACTER

It was considered the hill State's first smart city of Himachal Pradesh. As of the 2011 India census, Dharamshala had a population of 30,764. Males constitute 55% of the population and females 45%. Dharamshala has an average literacy rate of 77%, higher than the national average of 74.04%: male literacy is 80% and female literacy is 73%. In Dharamshala, 9% of the population is under 6 years of age. The number reached up to 53,553 in 2015, with a growth rate of 74.08% within five years. At present, Dharamshala MC comprises of 17 wards. Fig. 35 shows that all wards have nearly same population size.

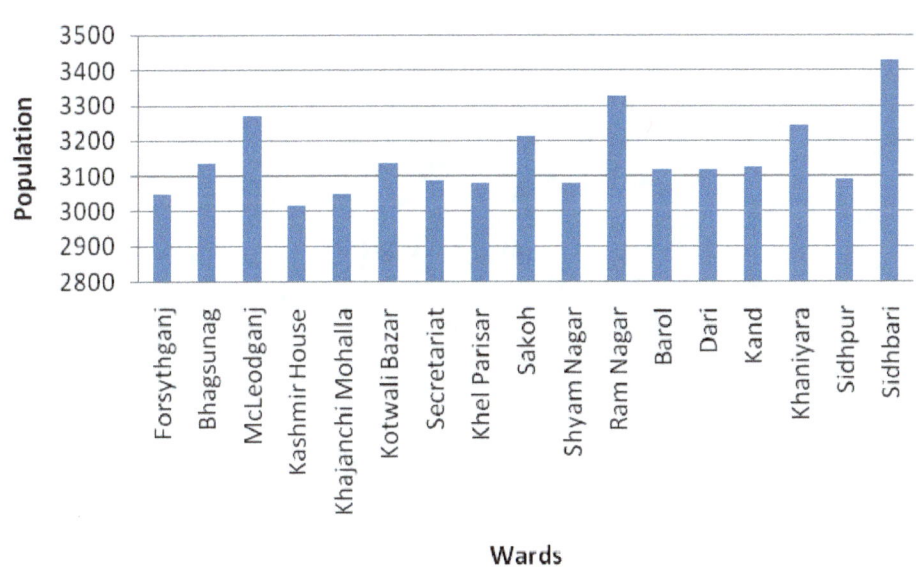

Figure 33: Demographic character of Dharamshala   Source: TCP department, H.P

3. GRAIN AND TEXTURE

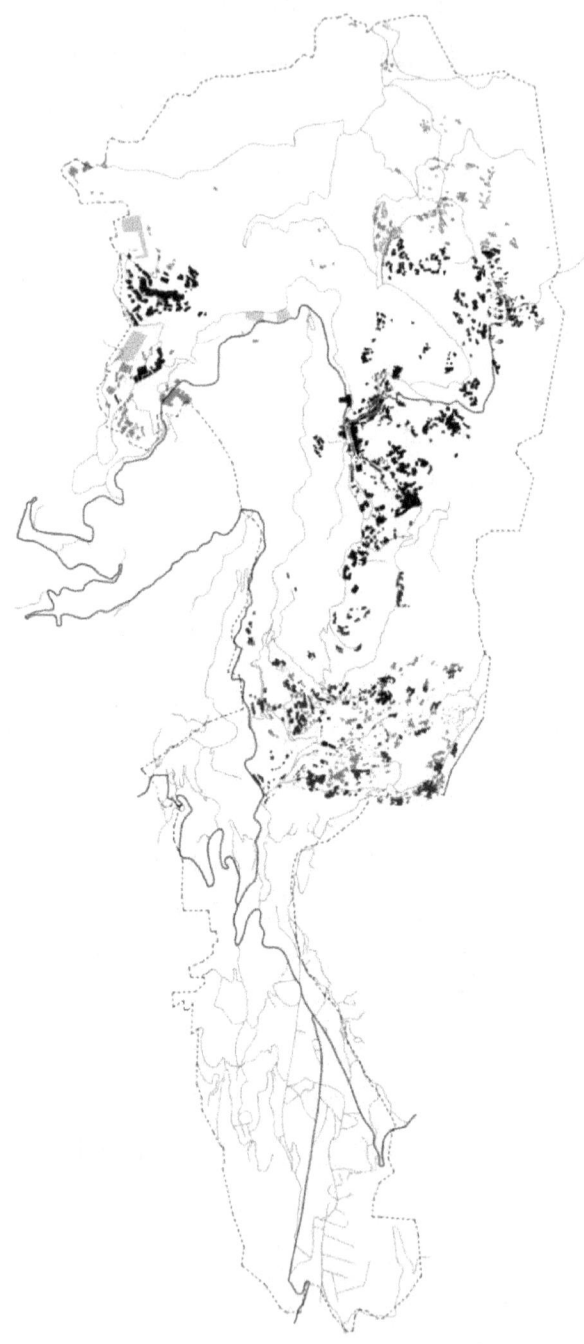

Figure 34: Figure ground map of Dharamshala

4. TRANSPORTATION AND MOVEMENT

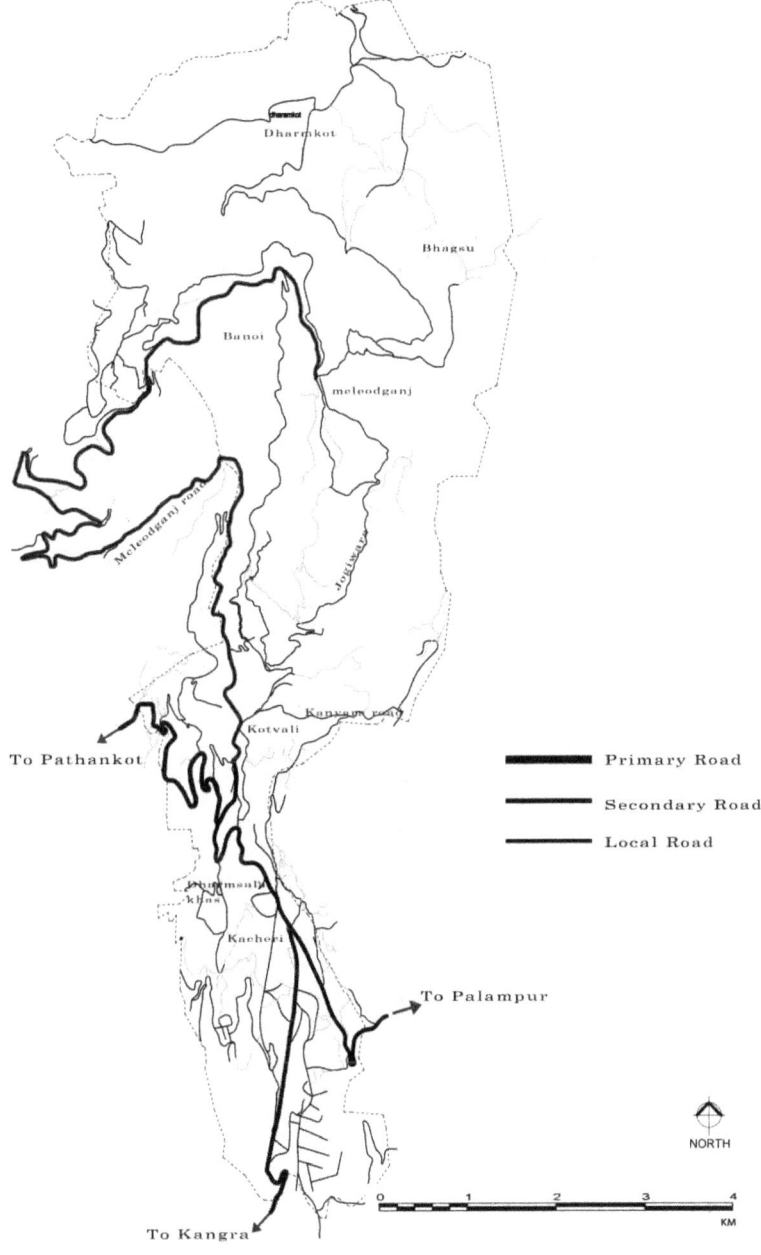

Figure 35: Main transit routes in Dharamshala

City has a strong tourism base with vast heritage of nature and culture. Dense tree layer and everlasting snow streams with and mesmerizing landscapes demarcates the Dharamshala city. City also has beautiful tea gardens. Kangra tea is very popular across India and the rest of the world. The city is dense along the vehicular roads with fine grain indicating lack of open spaces and pubic buildings. Fig. 36 indicates that due to unplanned development city has become very unsafe for disasters and need alternative connectors with activity centers.

Fig. 37 shows that the city has only limited transit corridors and it is dominated by narrow streets and pedestrian connectors. Lack in public transit and weak non-motorized transport (NMT) network creates inner city movement difficult. No organized space is available for the street vendors in the town. Hence, it is creating congestion near DC office, McLeodganj Sabji Mandi, Temple Road McLeod ganj, Dalai Lama Chowk, Near Bhagsunag Temple, Street vendors in Kotwali Bazar, Near Bus Stand, Near Zonal Hospital, Near DC Office, Opposite SP Residence, Near Government College, Dari Ground, Near Sacred Heart School, Near Shila Chowk and other areas of the town along major roads. Threats of the city also include unplanned growth and ineffective land management, huge floating populations, steep vehicular growth, change in climate and hazard vulnerability. These problems further degrade the aesthetic and potential of the area.

CITYSCAPE

1. LANDMARK AND ACTIVITY GENERATORS

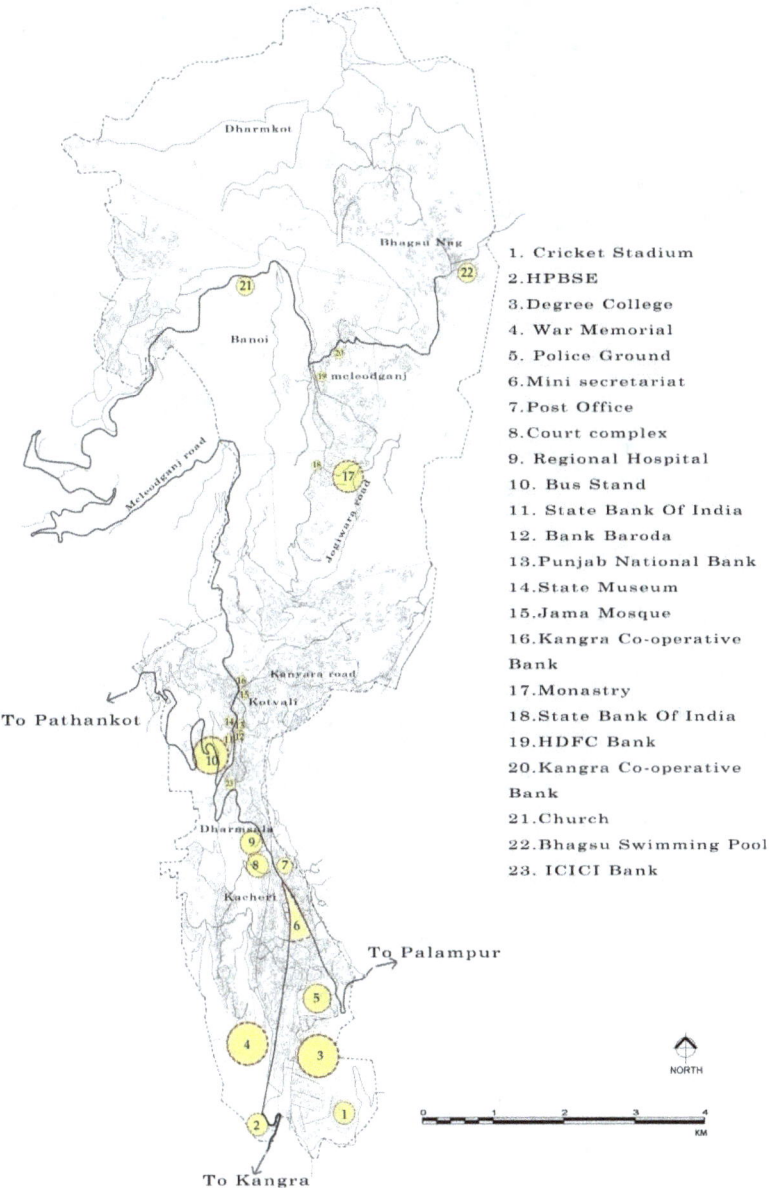

Figure 36: Major activity zones in Dharamshala

2. VERNACULAR ARCHITECTURAL VOCABULARY

Figure 37: Typical vernacular house near main town

Settlement has evolved over centuries in the form of organic spatial pattern and have inherent factors that deal with externalities like disasters and natural calamities. The houses utilize the slope and terrain for storage areas and plantation. Most of the houses face the south and east to derive direct sun. Houses are mostly linear and two storied, with kitchen and store on first floor and bedrooms on the ground floor. The built form does not hamper the natural terrain forms and hence reduces chances of landslides and instability. While we embrace newer technologies from across the globe, often we abandon or ignore the wisdom inherent in traditional practices.

3. CONTEMPORARY ARCHITECTURAL EXPRESSION

Figure 38: Modern RCC construction with sloping roof

Modern planned settlements and buildings are based on man-made scientific interventions and does not respond to calamities and climate change effectively. Transformation is happening are mostly due to changes in the lifestyle, peer competition, increased income, and social re-structuring. Ban on stone quarrying, high cost of wood and lack of trained masons are resulting in higher maintenance cost of the indigenous structures. Exposure to tourist activities and uncontrolled construction activities have changed the Image of the place.

CONCLUSION

Dharamshala is one of the fastest-growing urban centers. Nearly 1.5 Million tourists visit Dharamshala per year. City has a sustainable supply of energy, good health and education facilities, and civically aware community. But it lacks in infrastructure shortages to hold tourists. Over the years lack of adequate and efficient infrastructure in the city has led to a considerable reduction in the quality of life for citizens and tourists. Solid waste, wastewater management, and water supply are some of the major concerns.

Weak institutional and financial foundations of Municipal Corporation has also affected the trade of markets and their existence. Seasonal economy dependent on tourism and reduced employment opportunities is a major concern for local residents. Absence of initiative for research and development of traditional industrial products like shawls, cotton, and wool has impacted their quality, cost effectiveness and marketing.

Although the city is facing many problems but it has a potential to become a world travel destination. Places for leisure and fun through regional sports and adventure institutes can bring new life and job opportunities in city. Cultural economy and government land is available for the further development in the city.

OVERVIEW OF SUJANPUR TOWN

INTRODUCTION

Sujanpur, also known as Sujanpur Tira is a small town in the Hamirpur district Fig. 41. It was founded by Raja Abhay Chand of Katoch dynasty in 1748 A.D. The town is located on the bank of the Beas River. Sujanpur was previously

Figure 39: Location map of Hamirpur HP

inhabited by Maharaja Sansar Chand Katoch, the king of Kangra. He constructed his palaces, temples, and courts on top of a hill called Tira, hence the name of the town is Sujanpur Tira. In the middle of the town, there is one square kilometer of green ground, which in the local Pahari language is called the 'Chaugan'. Part of this ground is occupied by the Sanik School which was opened about three decades ago. The notable Holi fair also takes place on this ground, and lasts for almost three weeks during the month of March. Central core activity area is nearly about 3.03 km² with a density of 2,621/km².

1. CLIMATIC CONDITIONS

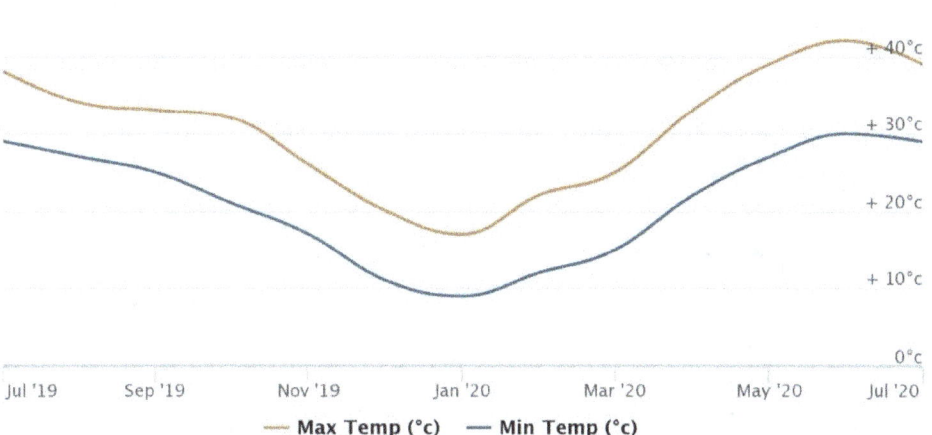

Figure 41: Average temperature in Sujanpur   Source: worldweatheronline.com

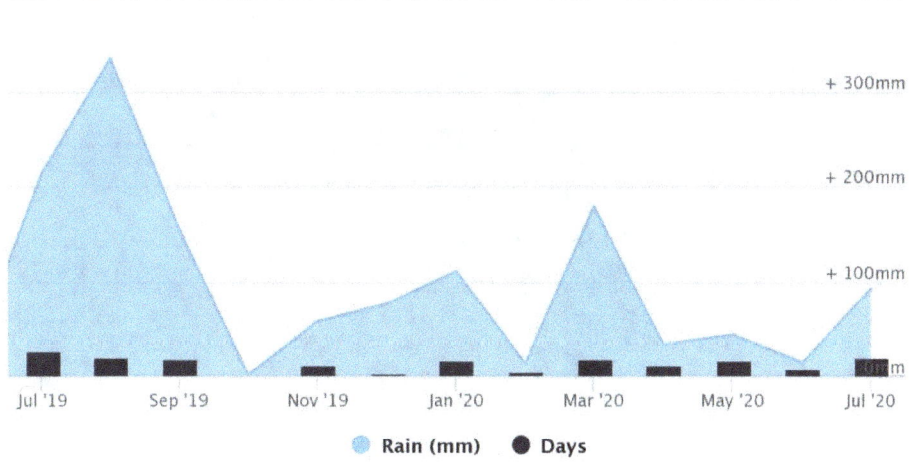

Figure 40: Average rainfall in Sujanpur   Source: worldweatheronline.com

The climate in Sujanpur is warm and temperate. Winter has much more rainfall compared to summers. In Sujanpur, the average annual temperature is 22.6 °C. The variation in annual temperature is around 19.2 °C. Fig. 42 and Fig. 43 shows the weather characyer of Sujanpur town. The average annual rainfall is 1977 mm. There is a difference of 589 mm of precipitation between the driest and wettest months.

2. TOPOGRAPHY

According to the planning commission of India any area above 600m in height from mean sea level or with average slope of 30 percent and above is classified as hilly shown in Fig. 44.

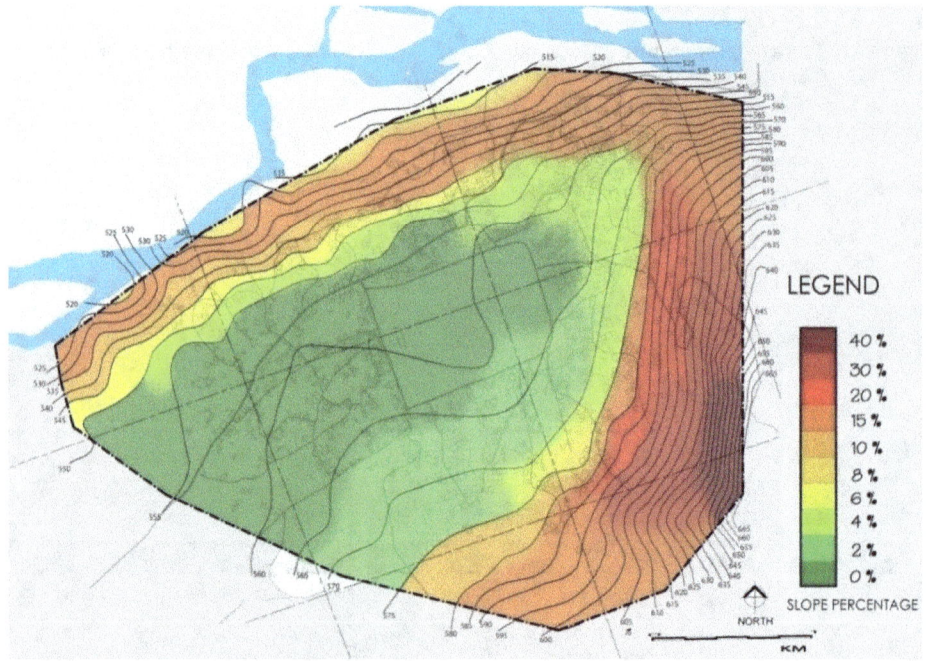

Figure 42: Contour map of Sujanpur

3. TOURIST INFLOW

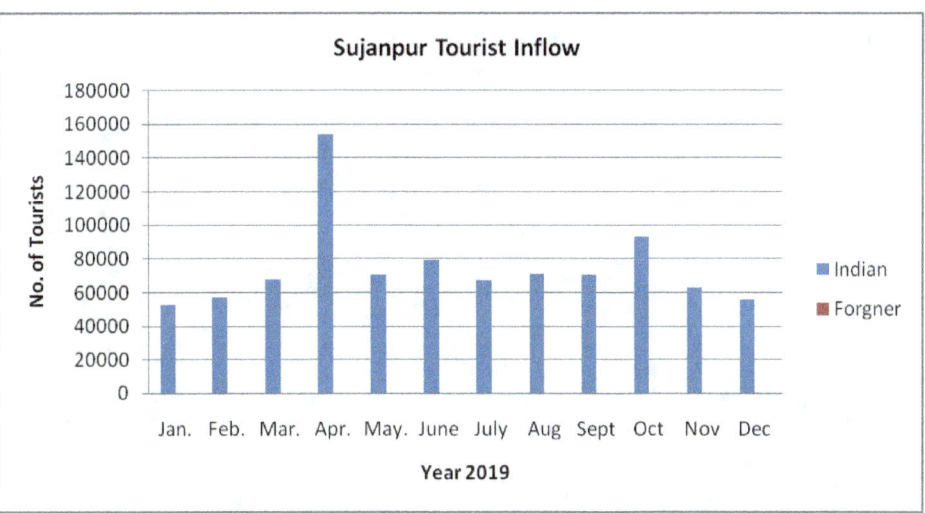

Figure 43: Tourist inflow per month   Source: Tourism department, H.P.

Sujanpur is a beautiful historical town. Sujanpur is the only city in the Hamirpur region which is open to tourists. This town is endowed with natural scenic beauty. In the recent years, it has attracted a large number of tourists from all over Himachal. Fig. 45 shows the monthly inflow of mainly local tourists in the town. Main tourist interaction is the fort. It is located on top of the hill and overlooks the Chaguan. Temples are another tourist attraction in the urban centre. They are mostly built by the kings of kangra. Some of the temples are places of great architecture and Kangra paintings. The Bansiwala temple, the Narvadeshwar temple, the Thakardwara, and some smaller temples are main tourist destinations.

MORPHOLOGY AND URBAN FORM

1. LAND USE AND GREEN COVER

Figure 44: Land use map of Sujanpur town

Total block area of Sujanpur is 184 km² including 180.88 km² rural area and 3.03 km² urban area. Fig. 46 shows the residential character of town. Most of the area is agriculture land and residential cluster can be seen in villages. There are 220 villages under Sujanpur block. The Chaugan is the biggest ground in the state. Sujanpur town is very well connected by road from all the major cities of Himachal Pradesh.

2. DEMOGRAPHIC CHARACTER

Tira Sujanpur has a population of 46,007 peoples. There are 10,693 houses in the sub-district. As of the 2011 Census of India the Tira Sujanpur Municipal Council has a population of 7,943, of which 4,262 are male, and 3,681 are female. The literacy rate of City is 91.73%, higher than the state average of 82.80%. Child (aged under 6 years) population of Tira Sujanpur (NP) nagar panchayat is 10%, among them 52% are boys and 48% are girls. Population density of the city is 2621 persons per km^2. There are 9 wards in the city; among them Tira Sujanpur. Demographic character of Sujanpur shown in Fig. 47 indicates that Ward No 07 is the most populous ward with population of 1172 and Tira Sujanpur Ward No 09 is the least populous ward with population of 693. There are 1769 households in the city and an average 4 persons in a family.

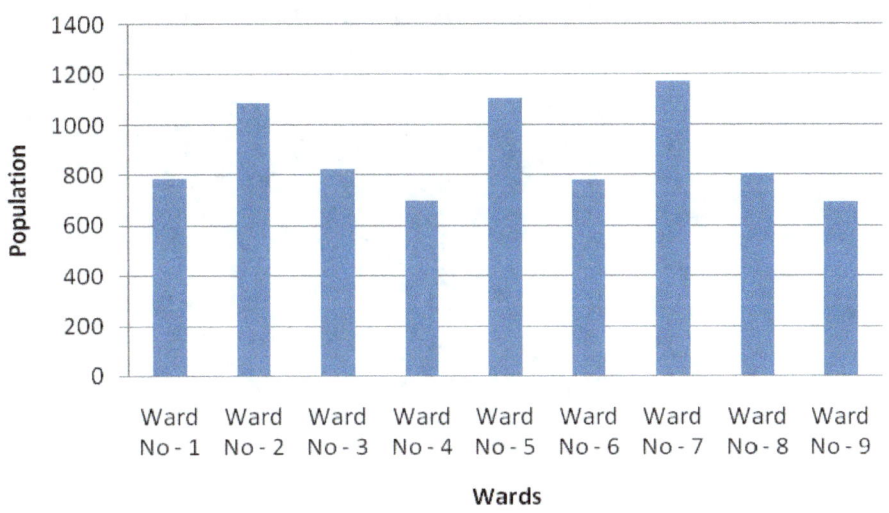

Figure 45: Demographic character of Sujanpur   Source: TCP department, H.P

3. GRAIN AND TEXTURE

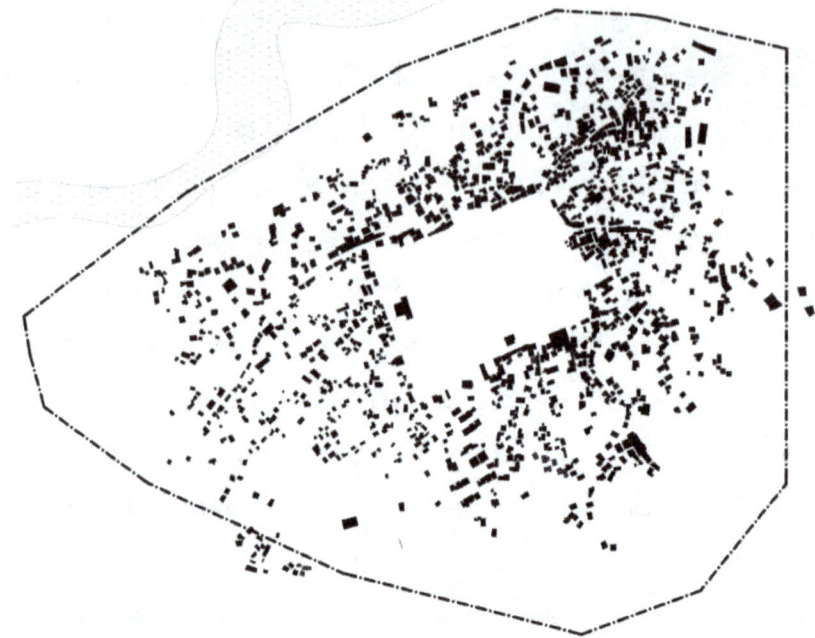

Figure 46: Figure ground map of Sujanpur

4. TRANSPORTATION AND MOVEMENT

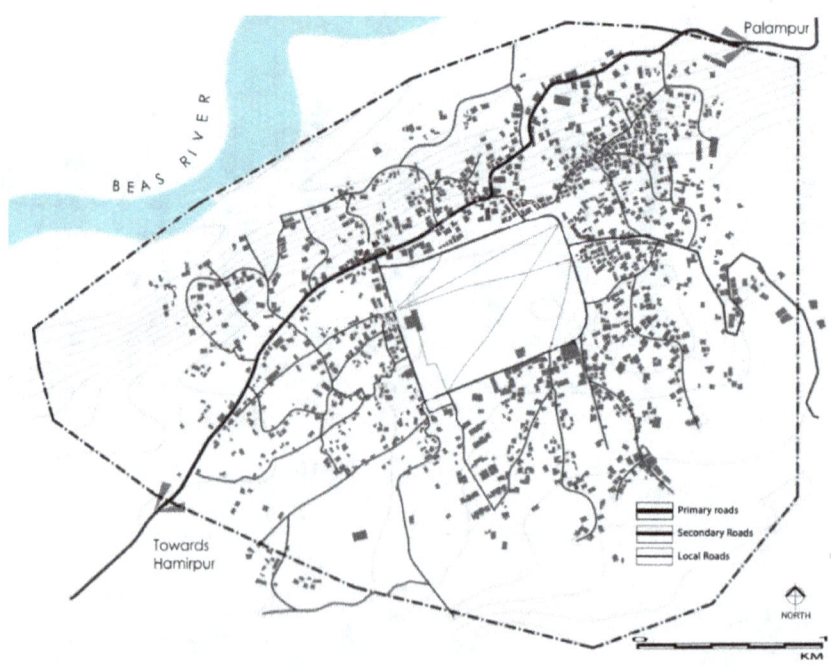

Figure 47: Main transit routes in Sujanpur

CITYSCAPE

1. LANDMARK AND ACTIVITY GENERATORS

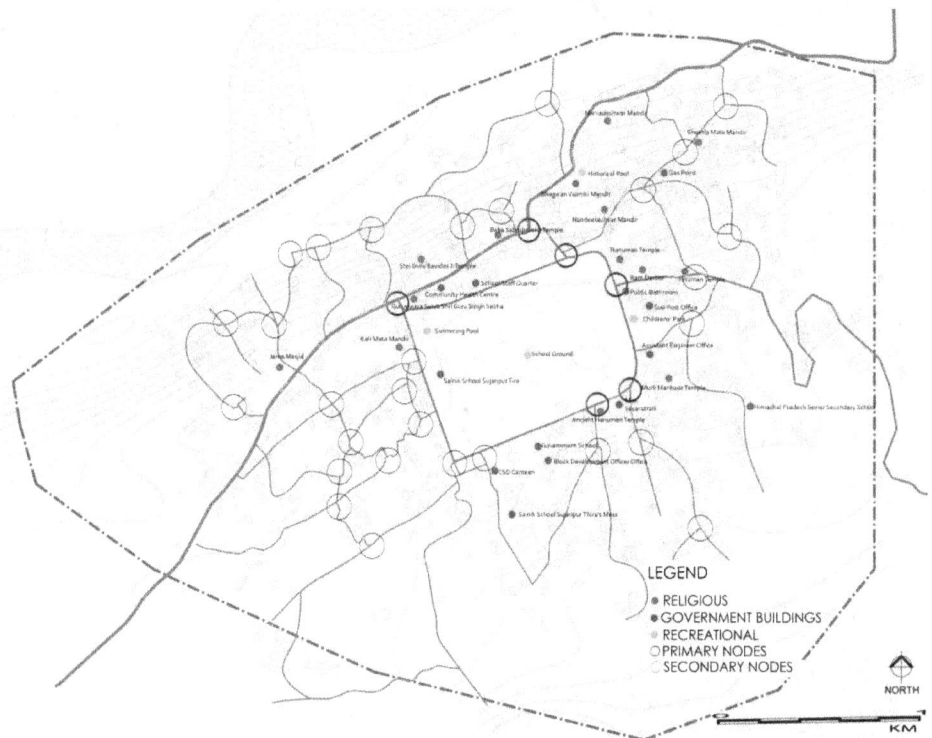

Figure 48: Major activity zones in Sujanpur

Maharaja Sansar Chand Katoch built large palaces and temples. The Murlimanohar temple stands on the left side of the ground, and the Narvadeshwar temple dedicated to Shiva-Parvti overlooks the bank of the Beas River. Another temple is Thakardwara. The temples were built in 1790 AD and 1823 AD by Katoch dynesty. Maharaja Sansar Chand also built the Chamunda Devi temple within the premises of the palace.

Figure 51: View of Chowki, Sujanpur Fort

Figure 52: Temples architecture of Sujanpur

2. VERNACULAR ARCHITECTURAL VOCABULARY

Figure 53: Typical vernacular house near main town

Figure 54: Village setting in hill town

3. CONTEMPORARY ARCHITECTURAL EXPRESSION

Figure 55: Modern street with RCC construction

The town is one of the smallest tourist destinations in hill state of Himachal. The typical character of low density, culturally rich and agriculture-based society are seen in this town. Most of the hill towns have this character. The main difference lies in the level of topographical differences which is evident from the study shown in previous city profiles. Fig. 51 and Fig. 52 shows old architectural marvels of the town. Like all big cities such small towns are also losing its character and under the economic forces all towns have become similar. These settlements lack urban planning and need timely interventions to preserve its ecological and cultural wealth.

CONCLUSION

In terms of settlement planning and design of building units in the context of extreme cold conditions of the hilly regions of Himalayas, it is imperative that we identify, document, understand and adopt the good practices and traditional wisdom. There is an urgent need to evolve the strong traditional knowledge base related to adaptation of architectural and planning practices. Urbanism and lifestyle demanding new typology of built masses. Land value, sprawl and market forces governing the form, size and shape the development. Lack of bye-laws, environmental awareness, and trained professionals is resulting in chaotic development leading to haphazard settlements. Contextual adaptation strategy for mitigating the adverse impact of climatic disasters and development of design strategies. Change in the mobility pattern, infrastructure, services and the transformation from rural to an urban character has not been addressed by the urban local bodies. The area contiguous to existing developed areas are anticipated to develop first due to availability of infrastructure facilities in the neighborhood and a gradual extension of these facilities to new areas is proposed to be ensured through public participation.

SYNTHESIS

This book is a small compilation of urban character of hill towns for planners and policy makers. The idea is to make them aware about a typical character which needs to be preserved for coming generations. Comprehensive framework and methodology are needed for carrying out future studies in hill regions. An acute shortage of funds for acquisition of land for public purposes is a great challenge to the administration to ensure a sustainable integrated development pattern. It has, therefore, been proposed to adopt a people's participatory approach for an optimum accomplishment.

For sustainable development in hill town's role of urban designers, environmentalists and community itself is critical. As a starting point this research sets a background for comprehensive planning and design approaches for urban and semi urban hill tourist hubs.

BIBLIOGRAPHY

1. A.K. Maitra, Development of Hill Capital: Shimla-2035, 52nd, National Town and Country Planning Conference on Development of Hill Capitals: Shimla vision 2025, Shimla, 2003, pp. 9–15.

2. Batra Adarsh, 2000, Himalayan ecotourism in Shimla, researchgate.net/publication/237780629_himalayan_ecotourism_in_shimla, accessed on 05.09.2015

3. Chattopadhyay, Basudha. 2008. Sustainable urban development in India: Some issues, Retrieved from: http://www.niua.org/Publications/discussion_paper/basudha_paper.pdf, 16th feb.2014

4. Clifton, K. Ewing, R. Knaap, G. J. Song Y. 2008, Quantitative analysis of urban form: a multidisciplinary review, Journal of Urbanism: International Research on Placemaking and Urban Sustainability, 1:1, p 17-45.

5. Das Pranab .2014, Mass Tourism & Environ- Infrastructural Crises of Shimla City: A Case Study, International Journal of Science and Research (IJSR), volume 3 Issue 11.

6. Dempsey N, Brown C, Raman S, Porta S, Jenks M, Jones C, Bramley G, 2010, Elements of Urban Form, Dimensions of the Sustainable Cities, Springer, London, 21-51.
7. Development plan Dharamshala planning area, kangra district, Himachal Pradesh, Town and Country planning department, Government of Himachal Pradesh. 2017.
8. Draft development plan, 2013, Town and country planning, Shimla, Himachal Pradesh.
9. Headman, R. Jaszewski, A. 1984, Fundamentals of urban design, American planning association, Planners press.
10. Isha Satmohini, Ray Srivastava and Chetan Vaidya, 2011, Planning for sustainable urban form for Indian cities , Urban India Journal, July-December issue.
11. Jenks, M. Burgess, R (eds) 2004, Compact Cities: Sustainable Urban Forms for Developing Countries, Spon Press. London, UK.
12. KPMG International, 2012, Tourism in Himachal Pradesh and the way ahead, PHD-KPMG in India report.
13. Kumar, Ashwani, Pushplata, 2013, Building regulations for environmental protection in Indian hill towns, International

Journal of Sustainable Built Environment, Volume 2, Issue 2, Pages 224–231.

14. Nag A. 2013, A study of tourism industry of Himachal Pradesh with special reference to ecotourism, Asia Pacific Journal of Marketing & Management Review, Vol.2, pp 89-106.

15. Nesamani, K.S. 2003. Sustainable transportation development in hill town- a case study of Darjeeling, Indian journal of transport management, vol.27, no.3, pp. 323-338

16. Ministry of Tourism, 2013, Tourism Survey for the State Of Himachal Pradesh, Government of India.

17. Ministry of Urban Development, 2015, Smart city mission transformation, Mission statement and guidelines, MUD, Government of India, Report.

18. Municipal Corporation Shimla, 2011, City sanitation plan for Shimla, GIZ ASEM report.

19. Sarkar, Amitava, 2011, Adaptive climatic responsive construction in High Altitude, World Academy of Science, Engineering and Technology International Journal of Civil, Environmental, Structural, Construction and Architectural Engineering Vol:5, No:12.

20. Sharma, P. Shree, V. 2012, Defining the impact of urban structure on energy consumption for developing sustainable city, Spandrel, Issue 5, pp47-51.
21. Sharma, P. Marwaha, B.M. 2014, Integration of Mobility pattern and cityscape for Sustainable urban form- Case of Shimla, ITMAR-October 20-21, Istanbul, Turkey.
22. Sharma, P. Marwaha, B.M. Dharmendra, Dr. 2015, Integration of new transport modes in urban form of hill towns for sustainable development, Indian institute of architects, Vol. 80, Issue 09, pp 38-42.
23. Sharma, P. Marwaha, B.M. Dharmendra, Dr. 2016, Investigating the Role of Multimodal Transport in smart city planning - Case of Shimla, SPANDREL 2015, Issue 11, Making Cities Smart and Competitive.
24. Sharma, Puneet, Urban design and Sustainable Transport in Hill Towns, Urban Rail Corridor Shimla, Glasstree Academic Publishing, 2018, ISBN 978-1-5342-0430-0
25. Shekhar, S. 2011. Urban Sprawl and other Spatial Planning Issues in Shimla, Himachal Pradesh, Institute of Town Planners, India Journal, vol. 8, no. 3, pp. 53 – 66.

26. Shekhar, S. 2014. Estimation of land surface temperature using landsat-7 ETM+ thermal infrared: a case of Shimla (Himachal Pradesh), Journal of institute of town planners, India, Vol. 7, no.2, 82-90.

27. Shree, V. Sharma, P. 2015, Tira Sujanpur: An Adobe of Katoch Dynasty, J. Inst. Eng. India Ser. A
DOI 10.1007/s40030-015-0111-5

28. Singh V, 2014, An Evaluation of Physical Performance of Himachal Road Transport Corporation (HRTC) in Himachal Pradesh, Confluence of Knowledge, Vol.2, Issue 4, October-December, 2014.

29. Sooden, M., Bist, N.S. 2007, District Human Development Report, Shimla, Planning Department, Himachal Pradesh.

30. Town and country planning, 2014. Shimla city development plan (SCDP), ILFS Report, India, Production Available online at http://jnnurm.nic.in/wp-content/uploads/ 2010/12/CDP_Shimla.pdf (Accessed on 28.07.2015)

ABOUT THE AUTHOR

He has nearly 15 years of experience in the field of Architecture and urban design– Teaching, Research and Profession. Currently working as Assistant Professor of Architecture in the Department of Architecture, National Institute of Technology, Hamirpur (HP). He has a Ph.D. Degree from National Institute of Technology, Hamirpur (HP); Master degree in Urban Design from School of Planning and Architecture, New Delhi; and Bachelor of Architecture Degree from Visvesvaraya Technological University, Belgaum, Karnataka.

Author has published research papers in refereed International Journals and Conferences. He has shared plate form with renowned researches at both International and national level. His current research interest is Urban design and resilience for sustainable urban form for hill towns. Author also works on community development activities for architects, engineers, faculties, and students through short term courses and expert lectures at many national forums. Varied consultancy and research projects in the field of architecture, landscape architecture, community resilience and planning have added to his credentials.

www.ingramcontent.com/pod-product-compliance
Lightning Source LLC
Chambersburg PA
CBHW070427220526
45466CB00004B/1574